Kant's Theory of Emotion

Kant's Theory of Emotion

Emotional Universalism

Diane Williamson

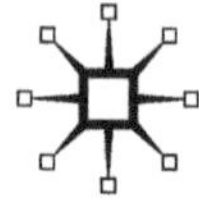

KANT'S THEORY OF EMOTION

Softcover reprint of the hardcover 1st edition 2015 978-1-137-49981-3

First published in 2015 by
PALGRAVE MACMILLAN®
in the United States—a division of St. Martin's Press LLC,
175 Fifth Avenue, New York, NY 10010.

Where this book is distributed in the UK, Europe and the rest of the world, this is by Palgrave Macmillan, a division of Macmillan Publishers Limited, registered in England, company number 785998, of Houndmills, Basingstoke, Hampshire RG21 6XS.

Palgrave Macmillan is the global academic imprint of the above companies and has companies and representatives throughout the world.

Palgrave® and Macmillan® are registered trademarks in the United States, the United Kingdom, Europe and other countries.

ISBN 978-1-349-50536-4 ISBN 978-1-137-49810-6 (eBook)
DOI 10.1057/9781137498106

Library of Congress Cataloging-in-Publication Data is available from the Library of Congress.

A catalogue record of the book is available from the British Library.

Design by Newgen Knowledge Works (P) Ltd., Chennai, India.

First edition: April 2015

10 9 8 7 6 5 4 3 2 1

Contents

Acknowledgments

Heartfelt gratitude is owed to those scholars, friends, and loved ones who helped me with this project along the way: Fred Beiser, Ray Boisvert, Robert Gressis, Samantha Matherne, and Kelly Sorensen; Laurie Marhoefer and Steph Clare; and Paul Hagenloh. Thank you to my dissertation committee, Greg Horowitz and Kelly Oliver, for showing early support for my ideas, and especially Jeffrey Tlumak for being a model philosopher.

Abbreviations for References to Kant's Works

A *Anthropology from a Pragmatic Point of View*. Trans. Victor Lyle Dowdell, ed. Hans R. Rudnik. Carbondale and Edwardsville: Southern Illinois Press, 1978. Page numbers in parentheses refer to the *Preussische Akademie der Wissenschaften* edition, volume 7.

CR *Critique of Pure Reason*. Trans. Norman Kemp Smith. New York: St. Martin's Press, 1965. Page numbers in parentheses refer to the original page numbers of the A and B editions.

CPrR *Critique of Practical Reason*. Trans. Werner S. Pluhar. Indianapolis: Hackett, 2002. Page numbers in parentheses refer to the *Preussische Akademie der Wissenschaften* edition, volume 5.

CJ *Critique of Judgment*. Trans. Werner S. Pluhar. Indianapolis: Hackett, 1987. Page numbers in parentheses refer to the *Preussische Akademie der Wissenschaften* edition, volume 5.

DR "Doctrine of Right" in *The Metaphysics of Morals*. Ed. and trans. Mary Gregor. Cambridge, UK: Cambridge University Press, 2003 (1996). Page numbers in parentheses refer to the *Preussische Akademie der Wissenschaften* edition, volume 6.

DV "Doctrine of Virtue" in *The Metaphysics of Morals*. Ed. and trans. Mary Gregor. Cambridge, UK: Cambridge University Press, 2003 (1996). Page numbers in parentheses refer to the *Preussische Akademie der Wissenschaften* edition, volume 6.

G *Grounding for the Metaphysics of Morals*. In *Kant's Ethical Philosophy*. Trans. James W. Ellington. Indianapolis: Hackett, 1994 (1983). Page numbers in parentheses refer to the *Preussische Akademie der Wissenschaften* edition, volume 4.

L *Lectures on Ethics*. Trans. Louis Infield. New York: Harper & Row, 1963.

O *Observations on the Feeling of the Beautiful and the Sublime.* Trans. John T. Goldthwait. Berkeley: University of California Press, 1960.

PP *Perpetual Peace: A Philosophical Sketch.* In *Kant: Political Writings.* Ed. Hans Reiss, trans. H. B. Nisbet. Cambridge, UK: Cambridge University Press, 1991 (1970).

R *Religion within the Boundaries of Mere Reason and Other Writings.* Ed. and trans. Allen Wood and George di Giovanni. Cambridge, UK: Cambridge University Press, 2004 (1998). Page numbers in parentheses refer to the *Preussische Akademie der Wissenschaften* edition, volume 6.

SF *The Contest of Faculties.* In *Kant: Political Writings.* Ed. Hans Reiss, trans. H. B. Nisbet. Cambridge, UK: Cambridge University Press, 19.

INTRODUCTION

Welcome to *Emotional Universalism*![1]

The goal of this book is to explain, defend, and apply Immanuel Kant's theory of emotion. When we come to understand Kant's explanation of the nature of emotion and the way we ought to morally evaluate emotions, we shall see that Kant's account can help us to make better sense out of our own personal, emotional experiences as well as evaluate contemporary culture and other theories of emotion. Remarkably, Kant—the supposed demonizer of emotion—has a theory of emotion that not only helps to bring even-handedness to current trends and debates within the study of emotion but also aids in achieving psychological clarity in evaluating and acting on our emotions. Kant's theory—believe it or not—is not terribly complex; it is rather intuitive if we already have some degree of proclivity to accept the validity of moral value (and Kant believes that everyone does). Hopefully, we can learn something about emotion from Kant, even coming to experience and evaluate emotions differently—more morally.

After briefly explaining Kant's theory of emotion and moral theory in this "Introduction," we will step backward and approach the topic from the perspective of an outsider. First (in chapter 1), before we start any explicit Kantian exegesis, we survey more general approaches to emotion; next, more specialized theories from psychology and philosophy (in chapter 2), and then (in chapter 3), some popular psychological and philosophical theories of the overlap between emotion and morality. Starting with general accounts of emotion and slowly moving toward a focus on Kant demonstrates the need for Kant in the more common ways of thinking about emotion. Kant's theory is then easier to understand and apply. Without this context, the importance of Kant's theory might be missed. Nevertheless, those readers (Kant scholars) who are already convinced of the importance of Kant might prefer to skip directly to chapter 4, where the more rigorous presentation of Kant's theory begins.

KANT'S THEORY OF EMOTION AND THE COMMON EMOTION TERMS USED IN THIS BOOK

"Emotion" and "feeling" are common and generic terms that mean different things to different people. They are the concepts this book explains by imbuing them with Kant's meaning.

Kant's theory of emotion must be amalgamated out of his comments about what he calls "affects," "feelings," and "moral feeling." Kant does not use any term that can be easily translated as "emotion." His term *Affect* (Kant uses the Latinate spelling) has sometimes been translated as "emotion" but this is not the best practice since affects for Kant are only one kind of feeling. By *Affect*—which I translate as "affect"—Kant refers to short-lived, immediate, rather intense emotional responses, such as surprise. Nevertheless, according to Kant, affects are a type of feeling (*Gefühl*), and this designation directs us to a much fuller theory of emotion.

In *Anthropology from a Pragmatic Point of View*, Kant discusses the "feeling of pleasure and displeasure" alongside the faculty of cognition and the faculty of desire. The key point is that there are different types of pleasures. Merely physical pleasures and pains count as feelings, for Kant, but there are also pleasurable and unpleasurable feelings that are caused intellectually. These types of feelings are weird—at least from the point of view of a folk distinction between mind and body—because they are physiological experiences (bodily pleasures and pains) that originate from mental events.[2] For example, if someone hears a new idea that helps her to make sense out of many topics about which she has been confused, she will smile, take a deep breath, maybe even bounce up and down a bit—these are all physical expressions of her intellectual pleasure. Her heart rate will increase, and other pains will become less noticeable. We might say that just as a peach tastes pleasurable and sweet, certain thoughts also *feel* a certain way by creating certain sensations. This is the locus of Kant's theory of emotion.

It would not do to say that emotion is only a physiological event or that it is only a mental event because neither by itself would count as an emotion. If, in our discussions, I sometimes highlight the mental dimension of an emotion or the physiological dimension of an emotion, remember that it is only a way of speaking and that because Kant defines emotions as intellectually (or mentally) caused feelings, they always include both a mental and a physical element. Additionally, usually the word "feeling" will be used as shorthand for intellectually caused feelings, but not other types of pleasure and pain. Accordingly, I often use the term "feeling" synonymously with "emotion."

We will discuss Kant's theory of "passion" as well as his theory of practical reason in general, but it is important to make a distinction between desires and emotions.[3] Many Kant scholars count Kant's remarks about "passion" as a part of his theory of emotion, but that is not the best way to proceed. By "passion" Kant means a relatively long-term action orientation; for example, a collector's passion for cars. "Passion" is nearly synonymous with "vice" for Kant. The experience of personal hatred or greed is closest to his mind when he refers to passions. It is true that pleasure and pain usually *cause* certain desires—the desire to seek pleasure and avoid pain—as do emotions, but we normally accept that emotion and desire (or wanting something) are not the same thing.[4] If we were to ask someone, "How do you feel?" and if she were to say, "I want to strangle him," we would most likely correct her by saying, "Okay, you're angry." Wanting a certain thing, like to strangle a person, might relate to feeling an emotion—anger—but it is not the same thing. Most of the scholarly work done so far on Kant's theory of emotion has focused on his theory of desire and moral motivation. To limit our understanding of Kant's theory of feeling to his theory of desire and moral motivation risks missing an explanation of the cause of feelings (and even of many desires) and the way to morally evaluate them. Understanding the causes of certain desires can also help us make better sense out of those desires as well as Kant's theory of moral motivation. Similarly, if we do not understand that some physical feelings come from thoughts, for Kant, we are stuck thinking that there is an unbridgeable gulf between reason and sensibility.[5]

Kant writes that when we contemplate the moral law, we feel respect. It matters little whether we say that the ensuing moral action was motivated by our *rational* comprehension of the moral law or by the *feeling* of respect because the rational comprehension and the feeling go hand in hand.[6] Sidestepping the topic of motivation, which has in any case been sufficiently covered in Kant scholarship, makes room for a psychological treatment of the evaluation of emotion and the cultivation of virtue.

As with passions, Kant tends to have a derogatory view of what he calls affects, holding that they are impetuous. Nevertheless, not all feelings fit his description of affects. The moral feeling of respect plays a central role in Kant's philosophy, but it is far from Kant's only example of an intellectually caused feeling.[7] As Sorensen has demonstrated, Kant describes a number of virtuous affects.[8] While some feelings might fade upon reflection because they are based on irrational thinking, other feelings are based on perfectly rational and moral thought.

It is not only necessary to act on the latter, but their presence in us is also an essential ingredient of our moral goodness. Feelings can vary according to the degree to which they are based on moral cognition as well as the degree to which they drive us toward immediate action (see figure 0.1). In general, the idea that emotions can come from many different kinds of thoughts sets us up to evaluate emotions in terms of their underlying intellectual causes. We can question whether perceptions are accurate, judgments are sound, concepts are based on enough experience, convictions are selfish, and so on. One overarching theme of this work is that emotions must be individually understood—we cannot assume that all emotions are immoral, irrational, useful, and so on. Unfortunately, all too often we either ignore or repress our feelings, which means that the thoughts go unevaluated. If they are moral impulses, they never get acted on; if they are selfish impulses, they never get challenged. Another overarching theme of Kant's theory of emotion and of this work is that emotional experience calls out for evaluation, and oftentimes the most relevant form of evaluation is moral evaluation.

Perhaps confusingly for the reader, in chapter 2, we will discuss theorists who use these terms, especially "affect" and "feeling," in different ways than Kant does. Affect theorists use the term "affect" to refer to a physiological response that follows from an external event, like being startled. They use the term "feeling" to refer to our subsequent consciousness of the physiological response. Neither difference

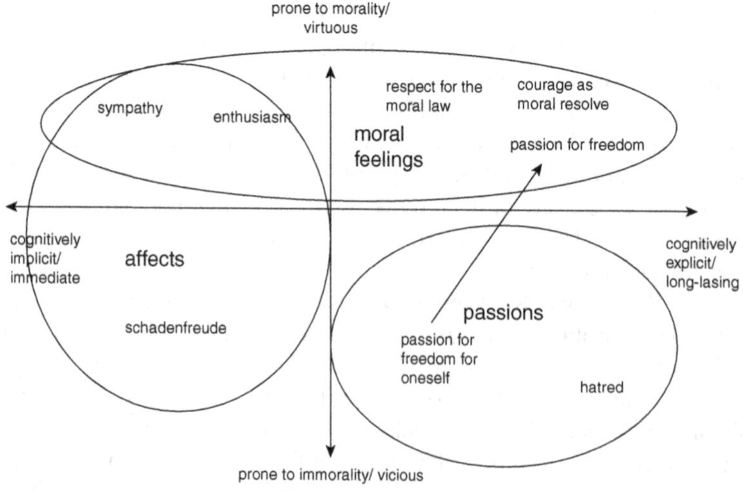

Figure 0.1 Cartesian plane.

in itself is necessarily consternating if we understand the distinction between affective theory and Kant's theory of emotion. Mostly, affective theorists hold that emotions are bodily responses ("affects") that are often a response to external stimuli. These bodily responses are then picked up on—felt—by the mind. While Kant would agree that there is likely a subsequent process of mentally making sense of the bodily sensation, he would class even perceptions as mental events, and so for Kant, the physiological occurrences come from mental events.

Most of Kant's comments about emotion are situated within his moral theory. Indeed, primarily his theory of emotion is a theory about moral emotional evaluation. The insight that feelings come from thoughts is very helpful for preparing us to then evaluate those thoughts and ask what we should do about them. Kant's ethics offers the next necessary step in evaluating our principles and making choices.

KANTIAN ETHICS

Kant begins his inquiry into ethics by asking whether or not there is something that is good in itself (G 393).[9] We know that there are a lot of things that are good *for* one purpose or another—Aristotle begins his investigations in his *Nichomachean Ethics* in much the same way—strength is useful if you want to lift a heavy box; a sharp knife is useful for cutting onions, and so on. Still, all of these things with mere instrumental goodness can also be bad if they are used in the wrong way: it would be better to be weak and without any sharp knives at all if one were plagued by murderous impulses, for example. One might easily conclude that there is nothing in the world that is good in itself—and that is quite possible. Nevertheless, Kant holds that this mere concept—the good in itself—is the starting point for understanding the meaning of morality.

"[G]ood in itself" describes something we do for its own sake, not for the benefit it brings, and morality is exactly the sphere of those choices that are chosen for their own sake.[10] Morality must be something that is good in itself because it is a consideration of the way that people ought to behave, not the way that they do behave. We cannot take a survey of human behavior, for example, in order to conclude that it is wrong to commit murder; instead, the justification that murder is wrong must be based in reasoning. Furthermore, moral justification must be based in objective reasoning. Moral laws are laws that, although they might not be externally compelled, as the laws of the government are, hold for everyone as determiners of action. If murder is morally wrong, it has to be wrong for a reason that holds for everyone. That is exactly what it means to say that something is morally wrong: it is wrong for anyone and everyone. Moral laws hold for everyone exactly because they are objective and rational. They do not change based on who is in power or what one might gain from breaking them. If something is morally wrong for *me* to do, it is wrong for everyone in that situation to do, and vice versa. For that reason, morality must refer to exactly those kinds of things that are good in themselves—or those things that should be chosen or not chosen simply because they are morally right or wrong. The reasoning for action that is based solely on moral considerations is called "a good will" (*ein guter Wille*) (G 393).[11]

From this analysis of the concept of morality, Kant derives some content about the types of things that are morally wrong and morally required. The moral law, which should govern all of our behavior, is "Act only according to that maxim whereby you can at the same time will that it should become a universal law" (G 421). To put this in plain English, it means: it is not fair to make an exception for yourself. It also means that you need to look at the "maxim" or the principle behind your action.[12] It might seem okay for you to take more than your fair share of something; if no one else does the same thing, then everything will work out fine. Nevertheless, it is still wrong if it is based on an inaccurate and biased principle—that your desires are more important than the desires of others. If your action is instead based on the principle of making sure that others have enough, then it is okay.

We can see that embedded in this moral rule is the notion that everyone counts: all people deserve equal respect and consideration. Kantian moral theory holds that humanity has inherent value. Similarly, Kant also formulates the moral law in terms of respecting humanity: "Act in such a way that you treat humanity, whether in your own person or in the person of another, always at the same time as an end and never simply as a means" (G 429).[13]

Most readers of Kant read only his *Groundwork for the Metaphysics of Morals* (published in 1785), wherein he makes this claim that a good will is the only thing that has inherent value, and they do not get to the *Metaphysics of Morals* itself (a much longer work, published in 1797), wherein he explains the notion of the good will. Consequently, many readers of Kant simply ignore the importance of the will, or moral consciousness, in his moral theory and focus instead of the different formulations of the moral law.[14] In the *Metaphysics of Morals*, Kant covers both those actions that are forbidden as well as the sorts of things that are morally required. Kant argues that it is our moral duty to strive for our own self-perfection and promote the happiness of others—*not* our own happiness and the perfection of others. Striving for our own self-perfection involves understanding what our moral duties are, and so it is the foundation for promoting the happiness of others. It is this comprehension of our moral duties and our commitment to fulfilling them that Kant means by having a "good will." While his wording in the *Groundwork* might confuse some—since he there states that a morally permissible action might be done for any reason at all—when discussing virtue in the *Metaphysics of Morals,* he clearly explains that we must not only do the right thing, but we must also do the right thing for the right reasons (DV 387). This subjective moral commitment, elaborated under the heading of the virtue of self-perfection, is the foundation for all morality, even justice.[15]

The term "will" (*Wille*) refers broadly to our entire capacity for practical reason,[16] not individual, specific choices, and so, by "good will," Kant is not referring to having "good intentions" even if something goes wrong with our choice. Instead he is referring to the entire mind, the entire faculty of reasoning about action, and the extent to which we are committed to making choices and adopting principles either based on what is morally good or based on what is (selfishly) good just for us.[17] It is that moral commitment that is at the heart of Kant's moral theory and good in itself. Kant further argues that it is because human beings have this amazing ability to comprehend morality, which is itself idealistic, and make goodness real in the world, that humanity has inherent worth; he calls this "dignity."

We do not, according to Kant, have a moral duty to promote our own happiness. Thinking that we have a moral duty to promote our own happiness might lead us to think that it is morally permissible to privilege ourselves over others. In any case, we naturally seek our own happiness, and Kant does not think it would be possible for someone to fail to take care of oneself. Instead, moral theory begins with the worry that people often exclusively take care of themselves to the detriment of others. Instead, "I ought to sacrifice a part of my welfare to others without hope of return because this is a duty." Nevertheless,

> it is impossible to assign determinate limits to the extent of this sacrifice. How far it extends depends, in large part, on what each person's true needs are in view of his sensibilities, and it must be left to each to decide for himself. (DV 393).

In other words, the extent to which you promote your own happiness, which is of course not prohibited, is between you and your conscience—it is merely not virtuous.

We are naturally adverse to sacrificing our own pleasure, and most people worry that fulfilling the requirements of morality will be unduly onerous. While no other person can tell you the extent to which you are called to sacrifice your own pleasure, I do believe that moral obligation is quite demanding. Kant's notion of virtue as a struggle implies that the more difficult the moral actions, the more amazing they are. We will most likely be better prepared to fulfill our moral obligations if we do not expect them to be fun—although making them fun is also a duty since we are thereby more likely to follow through on them. Kant is also right to point out that often virtue is its own reward because of the moral self-esteem it inspires.

While it requires some degree of self-sacrifice, Kant's moral theory is nowhere near an ascetic adulation of suffering.[18] It is never morally praiseworthy, for Kant, to debase oneself or to make the practice of virtue into self-torture.[19] Kant writes that self-perfection includes both natural and moral perfection. The former includes the duty to "make ourselves worthy of humanity by culture in general" (DV 392); in other words, all forms of educational and cultural advancement are means of improving ourselves as human beings.[20] While we do not have a duty to promote our own happiness, we should probably understand the duty to perfect ourselves as including overall physical and psychological health. Kant expresses the commandment to cultivate amoral self-perfection thusly: "Cultivate your powers of mind and body so that they are fit to realize any ends you might encounter" (DV 392). This is not unlike Aristotle's notion of virtue as an excellently functioning human being.

The duty to perfect ourselves morally (which includes comprehending all of our duties, both positive and negative) limits our duties to promote the happiness of others, but the duty to promote the happiness of others should probably be seen as limiting our duty to perfect ourselves in any completely amoral sense. Educating ourselves about what is right and wrong is more important than making other people happy, but making other people happy is more important than improving ourselves physically. It would seem morally perverse, I think, to choose enjoying an exquisite and novel meal, however much it might edify, over giving that meal to a starving person. Similarly, it seems impermissible to ever refuse to help someone when he or she asks us directly—this seems to follow from a simple application of the golden rule, that we would want the other to help us. Nevertheless, we must remember that our amoral self-perfection often has a moral dimension to it since it is impossible to be generous if we ourselves are not well off and taken care of (in every sense). Indeed, psychological health plays a necessary role in the fulfillment of our duties (and vice versa).[21] Negotiating one's own true needs along with helping others is a necessary task of moral decision making.

EMOTIONS AND MORALITY

Understanding, evaluating, and often more fully experiencing our emotions play a starring role in the development of moral consciousness, respect, and generosity. In choosing to do good simply because we know that it is good, we feel what Kant calls "moral feeling" (DV 387).[22] "Moral feeling" facilitates good actions. Many different feelings (both positive and negative) can follow from moral thoughts and actions, and they help us follow through on morality. Cultivating these feelings is a part of our duty to perfect ourselves.

Kant's theory of emotion, specifically, the important role it gives to moral feeling, helps us to see that emotions often help us recognize our moral duties in the first place—both those duties we have to others and the wellsprings of moral consciousness inside ourselves. Without a sensitivity to moral feeling, we are likely to ignore the moral dimension of our lives entirely. This is not to say that moral duty can be reduced to following feeling, but it is nevertheless too late for morality, as it were, if we are merely morally evaluating various nonmoral emotions. Virtue, with all of the many feelings it involves, like love, concern, worry, anger, pride, and so on, must be a complete and continuous disposition and not just an occasional mental exercise.

Oftentimes readers of Kant mistakenly take him to be arguing that emotions are the problem and reason is the solution. That is not the case. It would be more accurate to summarize him as saying that selfishness or vice is the problem and moral consciousness is the solution. Emotions occupy both sides of this moral divide: there are selfish and vicious emotions and there are moral emotions. Kant might think that there are quantitatively more selfish emotions than there are moral emotions, but that is only if he is considering a particularly selfish person. Just as Tolstoy famously remarks that each unhappy family is unhappy in its own way, whereas happy families are all happy in the same way (and Aristotle remarks about the myriad possibilities for going wrong compared to the unitary possibility of getting it right), there might be a greater variety of ways that we can deceive ourselves about our motivations—especially because we will then try to deceive ourselves about deceiving ourselves!—than there are ways that we can experience moral truth and goodness. Nevertheless, the lived experience of moral decision making will always be emotional.

In order to be virtuous we must know our motivations, and yet it is almost impossible, according to Kant, for us to know for sure that we are not deceiving ourselves:

> The depths of the human heart are unfathomable. Who knows himself well enough to say, when he feels the incentive to fulfill his duty, whether it proceeds entirely from the representation of the law or whether there are not many other sensible impulses contributing to it that look to one's advantage. (DV 447)[23]

This is not necessarily a problem but a task. Striving for self-understanding is therefore a ceaseless enterprise, but it is the foundation of moral commitment nonetheless. Thankfully, trying to understand ourselves is both morally and practically useful, as well as a part of our normal psychological functioning.

We saw above that Kant holds that emotions are feelings that are caused by thoughts (or other mental events). In the context of discussing emotion, the moral task is to make these thoughts explicit to ourselves and then to morally evaluate them. Sometimes we will find that they harbor selfishness; other times we will find that they express moral duties. After we morally evaluate our emotions—or the thinking behind them—we will have a better sense of how to respond to them; that is, whether we should take steps to overcome them or take steps to act on them. Whatever steps we will then need to take will not, and perhaps cannot, be covered in this book. Psychological research and acumen is then required. In seeking to change our emotions we cannot know ahead of time whether behavioral, cognitive, or pharmaceutical therapies might be most effective and respectful. When it is necessary to act on an emotion, we will also need psychological knowledge about the ways other people are likely to react and the strategies that are most likely to be effective. Nevertheless, if we know what we need to do and are committed to doing it, we will be able to discover the best way.[24] Where there is a (good) will, there is a way.

OVERVIEW

Chapter 1 prepares the reader for a discussion of the evaluation of emotion by calling to mind some common, simplistic ways we typically evaluate emotions. It presents three caricatured examples of common emotional comportments: stoicism (referring to the person who rejects all emotion), romanticism (referring to the person who accepts all emotion), and positive psychology (referring to the person who rejects negative emotions and accepts positive emotions). The reader is prompted to consider his or her typical response to emotions and is encouraged to realize that emotions cannot be judged wholesale and must be individually analyzed and evaluated. This chapter is philosophically lighter than those that follow so that the reader can be eased into the finer-grained distinctions between emotions.

Chapter 2 begins the philosophical discussion of the nature of emotion. After considering affective and cognitive approaches, I show that they share common ground and that the debate between competing theories hinges on pretheoretical biases. We examine the extent to which emotions are mental and physical, internal and external, susceptible to mental control, and related to moods. Overall, I aim to convince the reader of the variety and ubiquity of emotional experience. Kant's theory of emotion is reintroduced here.

Chapter 3 begins to consider the way we should evaluate our emotions. I canvass other theoretical attempts (e.g., neurological and evolutionary) to link emotion with morality, and suggest a more intimate connection by observing that emotions often themselves involve moral questions and moral evaluation. Furthermore, they are sometimes complicated and troubled, implying their own (negative or positive) value. I argue that emotions require rationalist moral evaluation and that this type of evaluation is often internal to the experience of emotion.

Chapter 4 focuses directly on Kant's theory of moral feelings and his explicit remarks about feelings like respect and sympathy that accompany his theory of virtue. I argue that the human experience of these feelings is far more vast than is often supposed—these feelings already play a starring role in our lives. Nevertheless, the experience of these feelings is complicated by our lack of understanding of our moral duties and a lack of courage and moral commitment. We have the duty to be open and sensitive to moral feeling.

Chapter 5 demonstrates the ways we should further evaluate all our emotions by more explicitly explaining Kant's moral theory. The first two formulations of the categorical imperative generate the twin demands of universalism and egalitarianism. Taken practically, the categorical imperative generates emotional duties to, on the first score, develop self-transparency and overcome selfishness and, on the second score, respect both ourselves and others. Morally speaking, we must realize that, as a human being, we have no more or less inherent value (dignity) than any other human. (Readers who wish to focus on Kant's moral theory are directed toward chapters 5 and 7.)

Readers who are more familiar with what is prevalently taken to be Kant's philosophy of emotion will be interested in chapter 6, which covers Kant's theory of affects and passions. I argue that Kant's considered opinion is that affects are not as problematic as he sometimes makes them sound and that there are moral affects. Kant makes an important distinction between affects and passions—the first are feelings and the second are a part of the faculty of desire—but part of the reason he often discusses them together is to demonstrate the importance of not letting affects, like anger, turn into passions, like hatred. Indeed, his theory of passion is at the heart of Kant's moral psychology of evil.

Chapter 7 discusses the notion of a morally committed life and Kant's perfectionism. It addresses the question of *how much* we must be committed to morality and *how much* we must try to do good. I do not try to sell morality in terms of self-interest or happiness, but I argue that the moral notion of self-esteem—I develop a two-tiered notion of self-esteem from Kant's theories of self-respect and moral desert—is a good antidote to the depressingly hollow and narcissistic notion of self-esteem often found in psychological and educational forums.

Chapter 8 relates Kant's theory of emotion to the psychological construct of emotional intelligence. Not only is this connection beneficial for moral philosophers and emotion psychologists who are unknowingly working in parallel, but it also expands the popular, problematic notion of emotional intelligence, helping to develop a more complex and useful theory of emotion and character. I argue that a useful notion of emotional intelligence must be based on a rationalist theory of emotion and must be informed by rationalist moral theory, or else it risks being overly simplistic and devaluing emotion.

WHAT DOES EMOTIONAL UNIVERSALISM MEAN?

Admittedly, "emotional universalism" is a strange term, but it can help us understand the major conclusions of this book. The term is felicitous because it can be taken in three different senses, all of which are important.

First, our discussion here leads to the realization that the human world is emotional. "Emotional universalism" means that emotions are everywhere; they are an important part of human life. In developing virtue, we must develop emotional understanding and courage. We must live with uncertainty and relish the positive as well as the negative.

People often deny emotionality because of the fear of being overwhelmed: "If I sit and comfort you about your dead mother, then I will have to sit and be sad about everything that is sad, and I will never get on with my life!" runs the worry. (Our hypereconomy has created a "life" that is more than producing what we need to have good relationships.) This fear of being overwhelmed is the same fear that causes us to deny our moral obligations: "If I care about where my food comes from, then I have to care about where my clothes come from, then I have to think about the effects of all of my actions, and if I stop and focus on all that I won't be able to eat or work or do anything, and I have to live!" The fear of being overwhelmed is not a genuine fear because it is based on the spurious notion that if I cannot be morally and psychologically perfect then I should not do one good thing or even try to improve. We must acknowledge what is good and bad, happy and sad; then we can try to be emotionally sensitive and do good. Developing a good character requires both that we be honest with ourselves, acknowledging and understanding the emotions of ourselves and others, and that we be patiently committed to morality.

In addition to accepting that the world is emotional and that certain emotional experiences are universal, we must look critically at our own emotions and emotional response. "Emotional universalism" also refers to emotional self-evaluation. Emotions have the ability to make us think that nothing else matters, especially when something else matters that we would rather not acknowledge. For example, let us imagine that a dear friend has suffered the loss of a close relative. Instead of visiting our friend in order to be comforting and helpful, we might feel "tired" or "cranky." We might suddenly feel that we never actually liked that friend as much as we thought. Kantian universalism offers us a method of working through and evaluating these emotions. It encourages self-transparency and the overcoming of selfishness. It also requires that we treat ourselves and others with respect and consideration.

Kant might sometimes imply that universality (the quality that our principles must have in order to be morally acceptable) is in conflict with subjectivity and that emotions are necessarily subjective.[25] This identification between emotionality and subjectivity is occasioned by a confusion about the meaning of the term "subjectivity." In one sense—the moral and intellectual sense—"subjective" is the opposite of rationally objective. Emotions can fall short of objectivity if they are selfish or based on flawed reasoning, but they need not be. In another sense—the phenomenological sense—"subjective" refers to any mental or physical process that belongs to a particular person. There is no moral problem implied by the fact that my experiences are mine, in fact the reproduction and application of moral concepts is not only subjective (in the second sense) but deeply intertwined with the experience of emotion.[26]

In the third sense of the term, our universalism must be emotional (because emotions are universal). Psychologically (and emotionally) there are primary (or universal) goods. Just as moral sense theory (from Hume, Hutcheson, or Smith, e.g.) can help us to understand that there is some sort of basic, natural moral comportment or moral sentiments, we must understand that people basically have the same needs and emotions. Following Kant, this commonality cannot be a foundation for prescriptive moral judgments. (We do not observe universality and prescribe universality. That move is theoretically barred.) Rather, we must realize, as Herman argues, that our moral evaluations require much practical knowledge and, I would add, specifically psychological and emotional knowledge.[27] To be emotionally blind and deny natural emotional needs is just as immoral as causing physical harm, and in many cases it is even more morally disgusting.

If we better understand emotion and morality, we can become more emotionally intelligent and virtuous. Still, it is not required that we acquire total knowledge about the content of our emotional and moral obligations before we start acting on them. We will have to build our ship while we are at sea. Instead, the conviction that there are moral and emotional duties and the commitment to try to make good on them can get us started. There is a good and a bad, a right and a wrong; we must continue to look for moral knowledge and do the best we can to live by it. These three insights of emotional universalism can start us off in the right direction.

1

Profiles of Emotionality

Emotions are simple—we are proud when we do something good, angry when others stand in our way—and yet, as humans we are infinitely complex in our ability to complicate them. While we all seem to have different approaches to emotion, there are, what we might call, common emotional types. We receive common messages about emotion from the media and our culture, so it makes sense that, in addition to normal emotional responses, we have similar approaches to emotionality itself. Psychologists might consider a list of emotion and mood disorders to be a list of emotional types, and, however true that may be, here we are looking for more general and common themes. For example, some people are very emotional and some people are unemotional. We shall see that each profile involves not only typical behaviors but also a set of beliefs about emotions and their value.

Here I depict three prototypical emotional comportments and speculate about their historical lineage, contemporary expressions, and natural consequences: (what I term) Stoicism, Romanticism, and Positive Psychology. Although you might, hopefully you will not see yourself in any of these types. The emotionality types we are discussing here are introductory caricatures. While I do discuss the real history of these movements, I am focusing on them as caricatures and not arguing that every person who identifies with these philosophies holds the beliefs I ascribe to them. I provide these caricatures as lenses through which to see the common messages we currently hear about emotions so that we can better evaluate them. I will provide contemporary examples of these types where I can, but it is likely the reader will be able to come up with some for herself. We can see that each of these types implies different assumptions about the value of emotion as well as the value of certain aspects of life. Illuminating these common psychological profiles will set the stage for a deeper consideration of the nature of emotion and its relationship to moral principles. The purpose here is to begin to think about what it means to have a general approach to emotion, uncovering our latent theories about emotion as well as our theories about value.

The point overall, perhaps, is that we should not have a *general* approach to emotion at all; different emotions are significantly different and each one requires understanding and evaluation. Furthermore, emotional evaluation cannot be simplistic. Evaluating our emotions requires a good deal of psychological acumen and moral deliberation, with which we will meet in the rest of the book. These three caricatured profiles of emotionality are all attractive, yet hazardous, shortcuts. While there is some discussion of Kant here, readers will not find rigorous philosophical analysis in this chapter, merely cultural observation. This chapter provides the jumping off point for the rest of the book by showing the relevance of and need for Kant's philosophy in our everyday lives.

STOICISM

First, we will consider the stoic person: someone who does not have emotions, or, if he does, he does not want to let others in on that fact. This style of emotionality might be accompanied by the belief that emotions are stupid, as it has been historically.

Historically, the Stoics were one of the most, if not the most, influential schools of philosophy. Zeno of Citium, the founder of Stoicism, began as a wandering and wayward soul. The philosophy of Socrates and his principal disciple, Plato, awed and unsettled the polytheism of the Mediterranean, and the world was ready for prophets. Zeno, before becoming the father of Stoicism, was attracted to Cynicism. This group of philosophers wanted to live in a way more in accord with nature. Diogenes of Sinope slept outside in a barrel and disdained all of society's artificial customs. Hipparchia of Marnoeia, who wore a toga instead of a dress and refused to live according to the traditional rules for women and marriage, chose to live with Crates of Thebes, who had given away all of his fortune, and study philosophy. "We're all just people," we might imagine Diogenes preaching; "Why does it matter what city I come from or what clothes I wear?" And so, wanting to challenge society's rules, the Cynics would go to the bathroom, have sex, and even masturbate in public; earning scorn and jeers, they were called "dogs"—*kynikos* is the Greek word for "dog-like"—a title with which they were perfectly happy. "After all," they might point out, "Socrates did not care about his appearance or his popularity. He only cared about the highest truths of reason."

Zeno, on the other hand, was not comfortable with the Cynic mode of behavior, and so Stoicism has been humorously described as "Cynicism for the shy."[1] Still, he was attracted to Cynic metaphysics: everything is guided by unwavering, divine, and rational laws. True freedom, he agreed, can only be found in becoming one with and following the edicts of nature. Yet, he broke away from this group and founded his own school of thought. Following Socrates and Diogenes, he believed that a philosopher should be an educative presence within the city, but turning away from both of their confrontational styles, he spoke from a front porch to whomever came to engage him. "Stoicism" comes from the Greek word *stoa*, for "porch." Following more faithfully on the insights from Cynic metaphysics, he argued that all ignorance and violence is also a part of nature; everything, good and bad, follows from Divine Reason. Therefore, it makes very little sense to try to make people change, as Diogenes had. Everything happens for a reason, and true peace can only be attained through understanding, acceptance, and detachment. The popular Christian serenity prayer—"God, grant me serenity to accept the things I cannot change, courage to change the things I can, and wisdom to know the difference"—was plagiarized from the Stoics. According to the Stoic sage, the only thing we can change is our own thinking.

Stoicism survived through three phases of revival into the Roman Empire, and Marcus Aurelius, the second-century emperor of Rome, was well known for his Stoic thinking and writing. The Stoic repression of emotion was perhaps more at home among the warrior-ethos of Rome than it had been in Athens, with its more recent memory of democracy. We might even say that the birth of Stoicism marked the takeover of the ancient Greek culture by the Spartan and eventually Roman cultures.

With regard to ethics and emotion, the Stoics prescribed apathy. Caring too much about any worldly, material thing is a recipe for unhappiness. Everything dies, everything passes away; peace and virtue can only come from accepting this eternal truth and going with the flow. Seneca affirms the following aphorism: "[The good man] yields to destiny and consoles himself by knowing that he is carried along with the universe."[2] Sherman accurately paraphrases the Stoic ethos as the injunction to "suck it up."[3] All emotions—happiness, sadness, anger—are the result of irrational, material attachments and false evaluations about the value of the transitory world as well as a failure to keep one's mind fixed on the universe's divine, eternal laws. Epictetus writes:

> If you love an earthen vessel, say it is an earthen vessel that you love; for when it has been broken, you will not be disturbed. When you are kissing your wife or child, say that it is a human being that you are kissing, for when the wife or child dies, you will not be disturbed.[4]

Indeed, in their quest for calm, the Stoics characterized emotion as a disease, a bodily disturbance. While they held that emotions (passions) are brought about by false opinions, they differed on whether or not original emotional disturbance was entirely controllable and eradicable. Even if one could never free oneself from the impulse to anger, for example, they all held that one should not act on emotional impulses.[5]

While Stoicism equates God with *logos* (the rational laws of the universe) and not the personal God of Christianity, Christianity took up Stoic asceticism easily enough. The Christian notion of the transcendent and immortal soul shares much with the Stoic sage's identification of the inner self as the true realm of freedom. Just as the Stoic asserts that even the prisoner can be free in his chains, orthodox Christians make ascetic sacrifices in order to purify the soul from the desires of the flesh. The fearless philosophy of Christian resignation—"Even though I walk through the valley of the shadow of death I will fear no evil" (Psalm 23:4 KJV)—might just as well be called Stoicism. The Christian adopts this Stoic stance because, believing in the immortality of the soul, the Christian should only care for eternal "treasures of heaven" and not "the works of the flesh" (Matthew 6:19–21 KJV; Galatians 5:19–21 KJV). Christian prayers and rituals remind their followers of the transience of the material world and the virtue of detachment. As they begin to purify themselves for their 40-day reenactment of Jesus's crucifixion (through which, we might add, Jesus remained almost perfectly stoic) and resurrection, Christians remind themselves that they are (nothing but) dust and to dust they shall return (Genesis 3:19). Christian dualism ("That which is born of the flesh is flesh, and that which is born of the Spirit is spirit"—John 3:6 KJV) is perhaps even more compatible with asceticism and apathy than Stoic materialism, and, in any case, we should not think that the Stoics have any sort of monopoly on being stoic.

The Stoic influence on Christian dogma can be seen more directly from Marcus Aurelius, through St. Ambrose, and finally to St. Augustine, a canonical church father. Seneca, Epictetus, and Aurelius were prominent figures from Roman Stoicism in second-century AD. During this same time, the early Christian Church was forming. Augustine was heavily influenced by Seneca, citing him alongside the Bible in his *Confessions*. Moreover, Augustine's work itself looks very much like a work of Stoicism, for example, with his rejection of grief at his mother's death, calling the expression of such a worldly attachment sinful.

Although it was more likely the case that Zeno was influenced by early Buddhism rather than the other way around, we can also see overlap between Stoicism and Buddhism. The position that suffering comes from attachment to the material world and enlightenment comes from detachment, or apathy, is the same in both. Therefore, if we accept this overlap between Stoicism and Christianity and Buddhism, at least in terms of apathy and asceticism, it is then relatively easy for me to assert that Stoicism is and has been the most widely accepted philosophy of emotion in the history of the world.

Indeed, there are still many signs of Stoicism in our culture. Quite often social norms prohibit the expression of emotion. "Being emotional" is thought to be a bad thing. Emotions are thought to be anathema to scientific pursuits and business transactions. There are still Stoics among us, or at least people who live by Stoic principles for some portion of their lives. When boys are told to "Be a man," they are being told not to express emotion. Indeed, Stoicism has always exalted masculinity for being rational and capable of enduring hardship. Femininity, on the other hand, is said to be emotional, and the female body is seen as more difficult to transcend. Stoicism prizes masculinity in all people—indeed, as we saw, masculinity itself is often defined in Stoic terms. Indeed, regardless of our depth of historical philosophical exegesis, the common English term "stoic" means "unemotional."

This denial of the importance of emotion is well entrenched in our commonplace understanding of emotion. The first thing one often learns about anger management from many a well-respected therapeutic source is to get rid of anger, supposedly, by taking a deep breath and counting to ten. Many emotions are characterized as a part of the "fight or flight" response that supposedly serves some purpose for irrational animals but with which we are for some reason still saddled. According to some, this response is a part of the "reptilian" rather than the "human" brain, although, of course, no human brain could function with its lower two-thirds missing.[6] For all of its technological successes, modern medicine knows very little about normal emotions, as it often relegates "mental health" to a separate, and all-too-often unrelated, field. Most of us imagine therapy as a place where people break down and cry, but just as often we understand therapy as the place where we can "get it out" and over with. For many of us, emotions are just a speed bump on the road of life.

Another example of the Stoic approach to emotions can be found in military culture.[7] This overlap makes sense because Stoicism was the dominant philosophy of the bellicose Romans; it came about in an environment of warriors and military generals, of gladiators and loyalty suicides. The current advertising slogan of the American Marine Corps, "an army of one," suggests Stoic elements: the Stoic warrior does not need emotional relationships. He has the sublime grasp of truth within, and this reliance on individual strength relieves those who do not accept the legitimacy of emotion. Boot camp and other hazing rituals are nothing if not Stoic training in ignoring feeling. You must become dead to your own pain—apathy is the goal—because you must see yourself as nothing but a soldier, a servant and messenger of the higher law. The body must become strong, clean, orderly, and uniform. All details about your personal feelings and attachments do not matter; everything is to be sacrificed to the higher ideal.

The Stoics often pointed to the deleterious social consequences of anger, and they promised a therapy that would help people live without their irrational passions. Nevertheless, Stoicism does not work: it itself often causes anger and irrational violence because the repression and denial of anger only leads to more intense anger.[8] Studies have demonstrated that it is nearly impossible to suppress thoughts. Emotional repression and suppression—"repression" is typically taken to refer to the complete denial of a thought or feeling as opposed to mere avoidance—causes a "rebound effect," which means that the thoughts and feelings actually become more frequent and intense. Similarly, thought suppression, such as Augustine's refusal to attend to negative thoughts about his mother's death, have also been shown to be associated with myriad psychological disorders, such as depression, anxiety disorders, borderline personality disorder, and post-traumatic stress disorder (PTSD).[9]

KANTIAN CRITICISMS OF STOICISM

Kant is often taken to be a Stoic, but this is an unfair characterization. He does write that virtue is characterized by apathy, but that is because he associates *pathos* entirely with selfish and short-sighted drives, *not* with feelings in general. While critical of moral sense theory, he does believe that there are important moral feelings that accompany virtue, and in agreement with Aristotle, he understands virtue as feeling the right feelings that follow from the right principles. That includes all types of feelings, righteous anger and noble sorrow, as well as justified pride and virtuous contentment. Most important, for Kant, is the feeling of respect, the respect for moral principles from which self-respect and the respect for all humanity is born. Respect might not seem like a feeling to many people, but more like a dry description of behaviors; on the contrary, it is a deep and profound feeling. Kant likens it to the feeling of the sublime, which is often accompanied by tears. Respect is a feeling of awe and amazement. It contains both joy and fear. It is the paradigmatic moral emotion, and, as such, it is the paradigmatic human emotion. If the stoic Marine Corps officer, for example, does not cry at a documentary about Martin Luther King, Jr., he has lost his humanity, for Kant, pure and simple. Although Kant often chastises desires for getting us into trouble (not acting on desires is not the same thing as repressing emotion), his philosophy prescribes worldly attachment: we have duties to ourselves and to other people, and these cannot be discharged without emotion.

Perhaps the most important point of difference between Kant and the Stoics is their theory of virtue: the Stoics held that virtue is sufficient for happiness. Kant, on the other hand, maintains a worldly and physical notion of happiness as the fulfillment of natural needs and desires. Similarly, he believes that this physical happiness is deserved by the virtuous person—it is the conjunction of these two ideas, virtue and happiness, that he calls the Idea of the Highest Good.[10] Kant does not identify virtue with asceticism. According to Kant, we have a duty to promote the happiness of others; "happiness" in this sense refers to their physical and psychological well-being. A Kantian ethic could never be identified with resignation and violence.

There seems to be an easy confusion between moral motivation and asceticism. (We see this especially with the role of women and the emphasis on sexual mores in more orthodox religions.) To be sensitive to whether or not one is being selfish or short sighted is not the same as rejecting the "pleasures of the flesh" wholesale, and Kant never suggests that such is necessary or desirable.

Fundamentally, the Stoics held that emotions are nothing but false judgments. We might agree that emotions are judgments and reject the idea that they are all necessarily false. We cay say, for example, that the feeling of sadness is the judgment that my father died and that it is a really horrible thing. Contrary to the Stoics, we might say, "Yes, it does actually really matter that my dad died; I should definitely feel sad." Nevertheless, it seems strange to many to posit that feelings are nothing other than judgments, especially because judgments seem to happen in the mind and feelings seem to happen in the body. We examine the way in which feelings involve judgments in the next chapter, but we should here note that Kant holds that both confused and clear-sighed mental states cause bodily feelings.

Stoicism has a wide appeal, and we should wonder at its attraction. It is true that many emotions are bothersome. Perhaps it seems, on the face of it, that it would be better to live without sadness or anger. Emotions are, in fact, based on our connection with the transient and mortal world. Many of the truths that emotions express make us uncomfortable; even those of love and happiness, for example, remind us of our own mortality. When we love someone we must face the horrible fact that that person will die, perhaps even before we do. The more we love, the more we must accept this vulnerability, sacrifice, and pain. Following the therapeutic philosophy of David Schnarch, we must note that in order to truly love anyone we must be able to look into the eyes of our loved one and say, "It's okay, you can go first, I will take the pain."[11] Without this acceptance of death and pain, we cannot fully engage in living.

As we shall discuss in the following two chapters, emotions involve their own (emotional) self-evaluation. In other words, we often evaluate our emotions while we are having them—and there is no reason to label the first evaluation as "emotional" and the second as "rational." For example, sadness about a friend's death can be simultaneously mellowed by concern for what will happen if we allow ourselves to experience sadness "too much." Anger can be accompanied by self-contempt and despair—the wish that we knew a better way to handle the situation. This self-contempt might then even cause more frustration and anger. Repression and avoidance are the *simplest* ways to make sense out of emotions. We have been taught to "make ourselves feel better" through diversions such as shopping, watching TV, drinking, and so on. Most of us have not been provided helpful emotional role models and so we do not in fact have any idea how to go about being sad or angry, or even happy, in a "normal," straightforward way. Plus, in a Stoic culture wherein emotions are devalued, they come to look feminine, "gay," or "dorky," and strength is seen as rising above emotion.

ROMANTICISM

Stoicism is an influential philosophy because it expresses a universal psychological tendency. Similarly, some Romantic elements can be found in all cultures, just as we see a beautiful myth illustrating the origin of romantic love in Plato's *Symposium*. Historically, the term refers to a backlash response to the "age of reason" in Europe in the eighteenth and nineteenth centuries, championing art, emotion, and nature. While not wholly unscientific, Romanticism strove to preserve the divinity of nature, the coherent whole of the universe, and a personal connection with "truth."

Above all, Romanticism champions emotion, all emotion—emotions are good. Emotion is thought to be a more authentic connection with reality than reason and is therefore capable of yielding more powerful truths. The ideal Romantic emotional comportment is to be in love—romantic love, with all of its ups and downs. The quintessential Romantic novel is Goethe's *The Sorrows of Young Werther*, wherein the titular character pines away in unrequited love, leading to his suicide. Although this emotion is painful, it is seen to be supremely authentic and important. In contrast to the Stoic philosophy of detachment, Romanticism is the philosophy of attachment. Romantic religious movements feature ecstatic unification with God and lurid expressions of the agony of carnal tortures, including the paradigmatic torture, isolation from God.

To the Romantic living among us today—the reader must remember that we are painting a caricature—all emotions are important. Sadness, anger, joy, sympathy are all important and all of these must be expressed. The Romantic values emotions simply because they are *hers*. They are the drama of life; they constitute who we are, our very identities. Finding personal identity and being true to it are important pursuits for the Romantic. The self is thought to have value simply because it is the self, and authenticity is a basic value.

In contemporary American culture, we more often embrace expression rather than suppression of extreme emotions. Reality television shows are a good example of emotional extremism. Similarly, consumerism defines our culture, and this pro-emotion stance of following our drives and passionately pursuing our desires usually involves buying something new. The Romantic prizes passion—passion for the new and exotic—but only because the new and exotic can themselves be the objects of passion; there is no shortage of products that can be dreamed up to embellish one's identity. The Romantic emotional comportment involves constant development of the new. (Habermas makes a similar point about modernity, or *Neuzeit*.[12]) Fashion trends, for example, always begin as an expression of uniqueness, and then, as they become popular, yield to the development of the next *new* trend. In turn, the constant barrage of the novel instills in us a Romantic preoccupation with our own whims and rewards impatience. (Historically, Romanticism is seen as the response to industrialization, but it seems that this newly narrow understanding and myopic pursuit of "fun" and "recreation" is impossible without the previous detachment from nature and surfeit of material wealth, as well as manufactured desire, created by industry.)

Ironically, just as the repressed emotion of Stoicism tends to rebound into even more emotion, Romanticism seems to develop into apathy and boredom. Flitting from one desire to the next does not feel very "authentic," and fixating on any specific aspect of one's identity often feels arbitrary. Hence, the Romantic type feels tired and washed up after awhile, bored by even herself. Nevertheless, new passions, especially love, are often found to cover over this sinking feeling of worthlessness.

In addition, it sometimes seems that the Romantic has trouble tapping into normal emotions. While the media teaches us that we should passionately pursue love, food, new cars, beautiful houses, and well-decorated nurseries, the emotional plight of others is presented dryly and disinterestedly. Politics rises to a fever pitch only when the government threatens to take away *our* freedoms; other than that, in our Romantic culture, we are politically apathetic. The Romantic also prizes spontaneity and originality, and yet many emotions reveal our rather normal and mundane attachments to ordinary things. Emotions are often simple and redundant; the most important ones tend to go on and on even after we have expressed them once or twice. The more important they become, the less dramatic they are.

Allowing ourselves to be thrown about from whim to whim, from reaction to reaction, is tiring, and the Romantic is left with no sense of what is genuinely important. Simply because we happen to have a certain emotional response does not make it good or worth having, and the Romantic is unable to discriminate between pursuits that are truly valuable and deserving of much energy and passion and those that are not. In short, the Romantic is mired in selfishness, and when the self has no foundation for value outside of itself, it collapses into itself.

We see this "emotions are good" assumption in many of the theories of emotion prevalent today. Affective theorists argue that emotions are physiological responses to certain stimuli that exist because they have evolved to be functional. Others argue that emotions give us important information about our values. Many highlight the role emotion plays in narratives of the self and identity. Nevertheless, this "information" and these "narratives" about value do not help us decide what we *should* value, and the functionality of evolution is not itself a foundation for value. We need a theory of emotion that can help us evaluate our emotions and find good responses to the challenges that they pose.

Why are we attracted to this simplistic "emotions are good" stance? The emotions do need to be defended; we do need to insist on our right to be emotional. Furthermore, as we saw above, denying emotion can cause psychological malady. Nevertheless, the basic fact of human emotionality does not itself make every emotion correct. It is easier to simply affirm every emotion—my feelings are right because I *feel* them—than it is to develop strong, deliberative reasoning skills. Evaluating emotion means sometimes admitting error: this is a difficult task. Additionally, without some acquaintance with moral principles, the emotions can themselves appear to be moral principles. Emotions do, in fact, sometimes anchor us in genuine moral commitments that we might otherwise ignore; indeed, there are some genuine moral emotions. Plus, all emotions feel, at least a little bit, like real moral commitments, and so, if we do not know how to evaluate our value judgments in any other way, we will be left thinking that the emotions are themselves moral values. This is not a stupid mistake, and it is common among theorists of emotion and morality even still.

KANTIAN CRITICISMS OF ROMANTICISM

According to Kant, feelings can be the effect of various functions of the cognitive faculty. To command cognition to follow feeling, when these feelings are actually caused by cognition, would be both absurd and difficult. While we might feel strong emotions or inclinations, we always intellectually assent to our behaviors, and hence only reason is able to justify our choices.

For Kant, it is the moral law, which is intellectually comprehended, that tells us what is good and bad. In a nutshell, we must treat each other and ourselves with respect and live in a way that is considerate, coordinated with other people, and (for many Kantians) environmentally sustainable. The purpose of life is to try to make the world a better place, and we do this in pursuing the happiness of others and our own moral improvement. There are many moral emotions that we should experience fully. We are morally obligated to care deeply about certain people and care at least a little bit about all people. Our duties to care require quite a bit of feeling (although perhaps even more commitment to follow-through when the feelings have gone). There are also many, less-noble emotions that simply add to the spice of life. Nevertheless, we also often experience selfish and disrespectful emotions. As we will see more in what follows (especially chapter 6), it is quite easy to get caught up in pursuits that take us away from what is really important in life. We must be able to discriminate, and the best way to be able to do that is to have a strong orientation in moral value that does not leave us thinking that whatever I feel goes.

POSITIVE PSYCHOLOGY

"Positive psychology" describes the movement within the discipline of psychology to understand and promote well-being. Over the past fifty years, there has been a boom in scientific endeavors generated toward the understanding of happiness. Since we all value happiness, this endeavor is inherently interesting. For example, we can ask, scientifically, whether or not money makes people happy. (The answer is that after one escapes poverty and makes about $75,000 a year, which, we might note, less than a quarter of the US population does, then, no, not really.[13] Money, as Aristotle noted more than 2,000 years ago, is a tool, not an end in itself; its usefulness depends solely on the way it is used.) Positive psychology may turn out to be useful for debunking some common, even harmful, myths about happiness. People in general become happier as they get older, for example; this information might serve as an antidote to our contemporary cult of youthfulness. Nevertheless, the science of positive psychology is young, and its popularity makes it especially vulnerable to popular prejudice and fads.

One such trend is a simplified approach to emotionality: the identification of happiness with having only positive feelings. Therefore, in calling the person who only wants to be happy, "The Positive Psychologist," I hope not to offend those researchers working critically in the field of happiness studies; I am instead providing a possible reason to be skeptical about the popularity of their field. In response to the Romantic's "emotions are good" approach, we have the converse position: "only good emotions." This emotional comportment is a purified hedonism; it attracts the pleasure-seeker and speaks to him, "Follow me to REAL, long-lasting pleasure." As the Romantic matures, and she tries to add some method to her madness, she is attracted to this comportment. She hopes that if she incorporates Stoicism, the denial of some of her emotions, she will be rewarded with the fulfillment that she originally sought. The pursuit of happiness, for its own part, is perilously close to Stoicism and risks falling back into it because it has not yet found the true ground of value, and losing touch with the negative emotions, it risks losing all of them.

Those who seek to have only good emotions are not thereby interested in moral goodness, but simply in the positive or pleasurable. Of course, the best way to avoid negative emotions is to live in ignorance about the moral problems of the world, but this self-incurred ignorance and immaturity is similarly immoral. Popular positive psychologists tend to assume that happiness and related positive emotions are the goal of life. Positive emotions feel good and negative emotions feel bad. Thereby we tend to get sucked into thinking that negative emotions are themselves the problem. In the pursuit of happiness, we end up wanting to avoid all information that might cause us to feel bad about ourselves or others. Like the Stoic, the positive psychologist wants to live "in the moment" and might similarly try to "meditate" oneself out of anger or sadness. Just as with Stoicism, there is a tendency to think that one can reason and relax herself out of negative emotions.

"What is the word for someone who is always happy and 'in the moment'"? a pseudonymous questioner wonders on an online question site. One answer: "enlightened." Regardless of the accuracy of this terminology, there exists the strong cultural tendency to think that happiness, defined as having only positive emotions, is the goal of life and is morally valuable. We can see this idea of being "in the moment" held up as a standard of value used to evaluate the acceptability of emotions. The preference for ever-changing, short-lived emotions tends toward repression because, as we shall see, emotions are often exactly those things that are not "in the moment," but in temporality, and hence keep us rooted in the past and future. Although the positive psychologists strive for worldly happiness, this quest for only positive emotions only promotes the devolution back into Stoicism.

This "always happy" type of emotional comportment is quite common for a number of reasons. Many people are uncomfortable with negative emotion. Especially in a large group setting, sadness or anger often disappear to a change of subject or attention to eating and drinking. (Hence we seek out the company of others when we want to hide from these emotions.) "Be positive" seems to be an edict of living socially, at least in our society, as workplace psychology steers us away from that which is "not productive." Similarly, as Barbara Ehrenreich notes, there is a popular misconception that experiencing negative emotions is physically unhealthy.[14] As with Romanticism, consumerism also thrives on this emotional comportment; just about everything you can buy is something that promises to make you happy.

In the midst of all of this excitement over finding happiness, we must ask "Is happiness really something we should be looking for?"

Just as a distant star appears brightest when we gaze to the side of it, perhaps happiness is something that can only be sought indirectly. (Philosophers sometimes call this likelihood the "hedonic paradox": we want to be happy, but seeking happiness cannot make us happy.)

Surely everyone wants to be happy, just as Aristotle notes in the very beginning of his famous *Nicomachean Ethics*, but what is happiness? Aristotle debunks the idea that happiness should be associated with pleasure, fun, relaxation, money, fame, or popularity. Happiness is a life well lived, a life lived excellently, or, what is the same thing for Aristotle, virtuously (with sufficient material wealth). A happy person is one who lives well, but in order to live well, we must know the purpose of life. Someone who plays chess well, for example, is not just someone who enjoys playing chess; he is someone who wins because that it the purpose of the game—to win. Since humans are the animals that can reason, to be an excellent human being, one must reason well, and reason well not just abstractly (about the profound truths of the universe) but also reason well morally and practically (which, for Aristotle, are the same thing).

It is difficult for positive psychologists to touch this depth of meaning. In asking people to rate their happiness, people most likely refer to pleasure. Researchers can ask people about "life satisfaction," but even this more refined notion does not fully get at what we are looking for.[15] If a truly vicious person reports a high degree of life satisfaction, for example, we would not want to count that person as happy, and we certainly would not want to learn from him or her how to live. In the absence of a moral foundation for our understanding of happiness, positive psychology all too often identifies happiness with pleasure or simply with the positive emotions.

Chris, Rob Lowe's character on the humorous NBC show *Parks and Recreation* exemplifies this "always happy," "Positive Psychology" emotional type.[16] He tirelessly pursues health and joy, but, of course, the viewer is quick to learn that, at the heart of it, he is depressed. Indeed, it does seem to be the case that the myopic pursuit of happiness goes hand in hand with depression. Sadness and anger are essential parts of life; they cannot be eliminated or ignored. When they are repressed, they only continually resurface in more pervasive ways. In addition, happiness is not the goal of life: goodness is. We all want to be happy, and we seek out those material conditions of happiness. Nevertheless, as Kant argues, it is only moral goodness that can make us *worthy* of happiness. Without moral goodness our "happiness" is just a bunch of stuff. He who is in pursuit of happiness senses this emptiness and has a gnawing feeling of his own lack of self-worth and of the meaninglessness of life. From the perspective of someone who prizes pleasure and personal fulfillment, life does indeed appear rather shallow and arbitrary: "I have become happy and fulfilled, but why do I exist at all?" In addition, for this type, depression preserves the truth of the negative emotions that have been denied. The voice of depression gains credence with its worry that perhaps sadness is the more authentic and honest engagement with the world.

KANTIAN CRITICISMS OF POSITIVE PSYCHOLOGY

We see in this case again that Kantian moral theory provides a clear lens through which to evaluate these popular modes of emotionality. First, as I noted earlier, Kant does not believe that happiness is the goal of life, but rather morally worthy happiness is the Highest Good. Unlike Aristotle, he tends to equate the term "happiness" with material well-being, so he thinks that it is possible to be "happy" and be evil. (Aristotle's use of the term does not permit that possibility.) An evil person would nevertheless not *deserve* to be happy, according to Kant. Still, it is possible to be a good person and suffer a bad fate. In fact, oftentimes the demands of morality come into conflict with our material well-being. Some pleasure comes from doing the right thing, even if there is no other reward or some great cost, but we still wish that good people would be rewarded with material happiness. We all, of course, look after our material well-being; that part of life is as inevitable as breathing, but that natural tendency is not the purpose of life. We must look instead to our moral duties, morally perfecting ourselves and promoting the happiness of others. Only then can we deserve to be happy, and only then does life have meaning.

Furthermore, for Kant, the myopic pursuit of happiness can end up looking like an addiction. His term "passion" refers to this kind of selfish, addictive drive. (We will learn more about Kant's notion of "passion" in chapter 6.) The exclusive pursuit of selfish inclination to the exclusion of any consideration of its moral permissibility and value is the height of evil for Kant, regardless of whether or not it makes one "happy."

The scientific findings of positive psychology are generally in agreement about the fact that human beings need to have good relationships in order to be happy. Yet, having "good relationships" does not just mean "being around people a lot." Our interpersonal relationships must be deep, loving, and honest. Emotions are the language of these relationships, and a good emotional comportment is one that helps us sustain good relationships, with others as well as with ourselves. This task is especially difficult when models of emotional health and a good theory of emotion are hard to come by. Our culture is filled with problematic messages about emotion and unhealthy models. (Sometimes it seems that television itself, with the preponderance of "reality," is becoming the genre of vice.) Similarly, everyone has spent some time either being stoic, romantic, excessively positive, or all three. Perhaps we have bounced around between them. The purpose of this chapter is not to point fingers but to help open our eyes to the possibility of a more intelligent, discriminating, and virtuous relationship to emotion. The tendency toward these emotional pitfalls is nearly universal, and simply understanding their attraction and danger is already a step toward self-understanding and moral goodness. That being said, we now move forward into a more serious consideration of the nature of emotion.

2

Understanding the Nature of Emotion

We seek an understanding of emotion and a better grasp of the role that emotions should play in our lives.[1] We aim to move past the wholesale evaluation of emotion as, for example, "useful" or "dangerous." In this chapter we will canvass two prevalent theories of emotion, affective and cognitive approaches, and see that they each privilege different types of emotionality as well as different types of emotions (as their main exemplars). Nevertheless, there are many different types of emotions and emotional experiences. Some emotions are simple, and yet some are very complicated. If we can say anything about the nature of emotions in general, it is perhaps that they involve some intersection between the mind and body. Kant's theory of emotion—as bodily feelings that are caused by various mental states (including perception)—helps to address a number of perplexities left by the current breach between theories. Not only does it show us the way that emotions are both mental and physical, it also helps us to make sense of the variety of emotional experiences. Moving beyond Kant, we will consider the ubiquitous nature of emotion and meditate on the question of whether or not we are always having an emotion. Similarly, we will note that, far from being "in the moment," emotions are often characterized by our connection to the past or future. While moral self-improvement is the larger goal, Kant urges emotional self-understanding as the necessary, although almost impossible, first step. Thinking about the nature of emotion and accepting its complexity can help us take that step.

What Is an Emotion?

The philosophy of emotion strives to answer this question. At first it looks like an easy question, but most people would hesitate to respond and prefer to be given the answer. As with Augustine's question, "What is time?" the subject matter evades direct observation. As soon as you look right at it, the present moment disappears before your eyes. Similarly, when you focus on an emotion, you might find a shortness of breath or racing thoughts, but you might still be confused about what is going on.

Wittgenstein warns us that not every question that makes grammatical sense also makes conceptual sense. That we can formulate the question, "What is an emotion?" does not mean that there is such a thing as "emotion." The question itself might be misleading: perhaps there is no *one* such thing as "emotion" at all. Indeed, those who study emotion disagree about whether or not "emotion" is a "natural kind" term or a "family resemblance" term.[2] Nevertheless, the position that emotions constitute a "natural kind," that is, that there is a relatively small set of definitive criteria for something being an emotion, is the more popular one. It makes sense that those engaged in offering a theory of emotion would like to say something of general and broad-sweeping importance about all emotions, not just something that holds true for some of the emotions some of the time. Still, that conclusion might be more of a disciplinary assumption than a falsifiable theory. In fact, most theorists of emotion simply assume that the question "What is an emotion?" has a single answer, rather than just a bunch of partial answers. Nevertheless, as we look at the answers that people have given to this question, we should remain open to the possibility that they are all right. Perhaps there are substantially different things that we can mean by "emotion," as well as substantially different ways that we should react to different emotions, and perhaps no one theory can explain them all.

AFFECTIVE THEORIES OF EMOTION

If we were to survey all of the theorists studying emotion now, we would find that William James's theory of emotion has been the most influential. Biology (including medicine) is perhaps the best represented discipline in emotion research, and James strove to provide the foundation for a biological consideration of emotion. He and James Lange are considered to be the founders of what is called the affective theory of emotion, which, as we shall see, does the best job of describing some manifestations of the emotion of fear.

In "What Is an Emotion?" James argues that emotions are the conscious recognition of bodily responses that follow from certain stimuli. His definition gives primacy to what he takes to be the first phase of an emotion experience: physical response, such as a rush of adrenaline. It is often believed that he means to reduce emotions to these physiological occurrences or "affects."[3] James famously quipped that we do not cry because we are sad, but we are sad because we cry. In other words, the emotion should be identified with the bodily event and not with the mental event that may accompany, follow, or even precede it. The thinking here is that if you are not experiencing your sadness physically in some way, then it does not make sense to say that you are sad. The sadness is just the physical affect.

Neurobiologists and neuropsychologists, following James, refer to the physiological response as the "emotion" and the conscious recognition of these bodily disturbances as the "feeling."[4] In this way they hold that they are studying emotion directly by bypassing the subjective, first-person account of emotional experience and by focusing on the parts of the body that react to external stimuli. Mental experiences are impossible to study in this way because we cannot observe consciousness. Following James, affective theorists hold that affects, or physiological responses, are a necessary element of any emotion. Without affects, they argue, we merely have thoughts, not emotions.

The affective approach seems to explain the emotion of fear, or startle, reasonably well. Many biologists believe that humans are genetically hardwired to fear certain objects, like spiders or snakes. Leda Cosmides and John Toobey, the directors of the Center for Evolutionary Psychology at the University of California, Santa Barbara, are the forerunners of the idea that we can understand much of the mind in terms of "psychological mechanisms," which are adaptations to solve biological problems that have a neurophysiological basis.[5] As with any mechanism, if a certain input goes in, a certain output necessarily comes out. Cosmides and Toobey take fear to be one of the most basic examples of a psychological mechanism. If this were true, then there would seem to be a direct neural circuit from the eyes to the limbic system, which is responsible for the rush of adrenaline of the fear response. Indeed, LeDoux's research on the fear response in rats seems to suggest that there is a direct route from the occipital lobes responsible for perception to the amygdala, causing the limbic response.[6]

James uses the metaphor of a key and lock to explain the biological fit and necessity of the connection between certain emotional stimuli and their emotion responses:

> Every living creature is in fact a sort of lock, whose wards and springs presuppose special forms of key, which keys however are not born

> attached to the locks, but are sure to be found in the world near by as life goes on. And the locks are indifferent to any but their own keys. The egg fails to fascinate the hound, the bird does not fear the precipice, the snake waxes not wroth at his kind, the deer cares nothing for the woman or the human babe.[7]

Here we can see the example of an adult response to an infant of the same species. For adult humans (perhaps a mother or a father), the example runs, the baby is like a key that unlocks certain behaviors in the adult, like cooing, holding, feeding, and so on. The strengths and weaknesses of this mode of explaining parenting behaviors are perhaps too obvious to point out, but it suggests to us the possible extent of evolutionary explanations. It is certainly true that for many of us, when we see a spider—or, perhaps more importantly, something that *looks like* a spider—we jump.[8]

Jesse Prinz attempts to develop an account of affective responses that does not rely solely on evolutionary cognitive mechanisms and can thereby bridge the gap between the automatic fear response, which is more akin to startle (what we would perhaps normally consider to be a neurological reflex and not an emotion), and other emotions. He merges a theory of "core relational themes," which are most likely innate in some way, and an idea of behaviorally conditioned "calibration files." So, for example, if I perceive that someone close to me died—Prinz insists that it is a perception—I will be sad. This is, perhaps, a natural response—a good candidate for a possible psychological mechanism. This specific perception (of a corpse, perhaps, or of an empty chair) leads to this specific emotional response (crying or depressed autonomic functions) because this event necessarily, most likely because of our biological evolution, has an emotional meaning; in terms Prinz borrows from Richard Lazarus, it triggers a "core relational theme."[9] This response is probably universal, and hence lends itself easily to Prinz's biological explanation, but another example, like falling in love, will require that we bring in Prinz's new idea of "calibration files."[10]

An example is perhaps warranted to explain the notion of a calibration file. In the case of falling in love, it is likely that there is a certain type of person and certain events that trigger the feelings of love. What we colloquially refer to as one's "type" has been conditioned by one's upbringing: talking to medium-height, dark-haired men about intellectual ideas stimulates the feelings of romantic love in me because my father was of medium height, had dark hair, and because my culture and my parents, perhaps indirectly, have instilled in me a sense of the importance of intellectual agreement. (I also love furry beards—but perhaps that's universal.) A totally different woman living in Beijing, for example, might feel the feelings of love when she comes into contact with a pale-skinned, broad-faced, religious man who owns a car; specific perceptions will lead to feelings of love in her because of her upbringing and her culture. These environmental conditions that are specific to the individual are what Prinz calls "calibration files," which combine with the universal core relational themes (love as the experience of finding a suitable mate, perhaps) to determine the affects that are triggered in us by certain events.

Prinz's goal is to explain the way that emotional responses are mostly reactive and outside of conscious control. Nevertheless, according to this account, emotions can sometimes be triggered by thoughts. Imagination is one thing—a dream about an intellectual, bearded man—but emotions can also come from reason. For example, I can wager that this man would be even more perfect for me if he liked children and were committed to being a parent (even more so than any of the men I had heretofore experienced). Prinz or Lazarus are likely to try to continue to explain the emotion—my amazement that this man is perfect for me—in terms of the *perception* (or the memory of the perception) of a core relational theme.

Similarly, thoughts can trigger the same physiological responses as real external stimuli do. For example, I have a rather powerful emotion just thinking about my children dying. On the affective account, these thought-triggered reactions are "as-if loops" because they unfold "as-if" there were some real external stimuli present.[11] Emotional disorders, according to an account that goes back to Descartes, occur when the mentally triggered emotions come to have a life of their own, detached from "natural" emotional stimuli. The important thing to notice is that, while the affective account acknowledges that some emotions seem to be caused by thoughts—imagination or reason—it holds that emotions triggered directly by external stimuli are somehow more primary or natural.

One of the most convincing criticisms of the affective approach is the retort that affect is not enough; mere feelings are insufficient to tell us which emotion we are having or if we are having an emotion at all and not just some other physiological experience.[12] In 1962, Schachter and Singer performed an experiment wherein they told subjects that they were receiving vitamin injections to study vision. In fact, the subjects were either injected with epinephrine or a placebo. Of those injected with epinephrine, some had the physiological side effects (racing heart, shakiness, etc.) explained to them, others did not, and a third group were misinformed about side effects (to control for general anxiety about the side effects). Participants were then left in a room for 20 minutes with a (fake) co-patient who was either happy or angry. Those subjects who were either given a placebo or informed about the real possible side effects of the injection were less likely to mimic the emotions of their co-patient. Those given epinephrine on the sly, on the other hand, were more likely to attribute their arousal to their situation and therefore to also report feeling either more happy or angry, as their co-patient had. The subjects needed a way to make sense of their physiological experiences, which were themselves indistinct, and they did so either by means of understanding them to be a side effect or understanding them to be an emotion relevant to the situation. Singer and Schachter believe that this experiment shows that physiological symptoms are not themselves sufficient for specifying an emotion.[13]

Affective theorists do not seem fazed by this criticism, arguing that it does nothing to change the fact that the affect is the essential core of the emotion: there would be no emotion without affects, they say. Prinz argues that the affective theorist can take the cognitive context of the affect into account, but that the affect is still the core of the emotion.[14] One might say that the scientist can peel away the external layers and study the core of the emotion (in the body).[15] Actually, affective theorists like James and Prinz are interested in the connection between external stimuli and internal physiological responses. It is this connection that they believe specifies the nature of the emotion.

Affective theorists are also correct to point out that the affect might have a life of its own that is entirely disconnected from conscious thoughts. For example, while catching my young child and preventing a fall, I will probably have a surge of adrenaline. Prinz counts the fact that I may remain a little jumpy—or my adrenaline might be channeled into excitement—even after the event is over as evidence that affects have a life of their own and should not, therefore, be linked to judgments. We might say that the antecedent (efficient) cause of the emotion might be different than the formal or material cause. While my anxiety is *about* the worry that my child could have fallen (formal cause; core relational theme) and started when I perceived him slip (efficient cause), the affect itself is bodily (material cause) and therefore incorporates everything else that is going on in my body—if I drank too much coffee, for example. It should seem strange that I am so excited about whatever comes next, and I should be skeptical about the authenticity and appropriateness of this emotion just on these grounds alone. Any quest for self-understanding will take these other possible material causes of emotion seriously.

To recap: affective theorists focus on the physiological events (affects) that follow from certain stimuli, and they tend to explain the connection between the stimuli and the affects biologically.

COGNITIVE THEORIES OF EMOTION

Cognitive approaches to emotion, in contrast to affective theories, take the mental aspect of the emotion as central. Similarly, they do not privilege external but internal causes, like judgments. Aristotle and the Stoics are taken as forerunners of the cognitive theory of emotion: Aristotle held that the emotions are based on beliefs, and the Stoics held that the emotions are nothing but judgments. The Stoic theory of emotion is simpler than Aristotle's, and there is more reason for calling it a cognitive theory, but Aristotle's theory is more intuitive and, like Kant's, helps us to bridge the gap between cognitive and affective theories.[16]

An Aristotelian theory of emotion comes mostly from Aristotle's *Rhetoric* and *Nichomachean Ethics.* In the former, Aristotle outlines both the subjective and objective causes of certain emotions. His emphasis is on teaching lawyers to sway jurors in ways conducive to their cases, and hence he highlights the ways that emotions are stirred by beliefs.[17] Drawing from Aristotle's discussion of emotion in the *Rhetoric*, Cooper argues that, for Aristotle, emotions involve three elements: a feeling that is either pleasurable or painful (or both), beliefs that arise "from ways events or conditions strike the one affected," and a desire or plan for some kind of action.[18] This definition of an emotion combines aspects of the affective, cognitive, and conative orientations of the mind.[19] Still, it is not clear from the *Rhetoric* whether beliefs or affects are more central to emotional experience. Aristotle writes: "Emotions are things through which, being turned around, people change in their judgments" (1378a24–27). Aristotle emphasizes the ways in which emotions and beliefs are related, and changing one can change the other.

A quick review of Aristotle's ethics will remind us that he believes that in some ways the emotions are susceptible to discursive reason and in some ways they are not.[20] The *Nichomachen Ethics* paints practical virtue as a prerational disposition. Aristotle states that the virtuous person does not have to *control* his feelings because he feels pleasure and pain at the right things. Aristotle's tripartite theory of the soul holds that emotions originate in the nonrational soul and must be educated by reason. This picture of the soul makes us question whether or not we should attribute a cognitive theory of emotion to Aristotle at all since, on this model, emotions are usually nonrational desires.[21] Then again, the very fact that Aristotle holds that emotions are susceptible to rational argumentation and evidence makes us wonder why he places them in the nonrational/animalistic soul in the first place. This classification is perhaps because his account of *akrasia* rejects Plato's assumption that total rational convincing of the passions is possible. In other words, Aristotle holds that real internal conflicts can exist between desire and reason; desire can overpower even the best will unless it has first been properly trained. For Aristotle, the emotions will always remain a bit outside of rational control; instead, they are more susceptible to behavioral means of modification, like mimicry.

As with all ancient and Hellenistic theorists of virtue, Aristotle sees virtue as a process of training. Sometimes training the emotions happens directly, and we would then say that the emotions are susceptible to reason, or it can sometimes only happen indirectly, and in these cases we would say that the emotions are not susceptible to reason. Furthermore, Aristotle holds that the latter kind of training (behavioral training), which must occur continuously through proper modeling and practice from infancy on, is the foundation for the former kind of training (rational alteration). Indeed, because Aristotle is primarily a nominalist and an empiricist, he does not believe that someone can even come to know the meaning of "good" or "bad" or how to properly respond to emotional situations unless he or she has been raised correctly (hence the importance of education).

The Stoics present us with the most extreme form of a cognitive theory of emotion: they held that emotions—or those passions that appear to overtake reason, chiefly fear, lust, pain, and pleasure—are simply beliefs, and they claimed that our prejudice that emotions are more than beliefs is simply a mistake.[22] Following Aristotle, they note that when you change your beliefs you change your emotions. Their focus is on being able to control oneself completely. Stoicism, as we have seen, shares similarities with a Buddhist philosophy of emotion and was very influential for the development of Christianity, especially through St. Augustine, who is arguably the most important historical source for the popular contemporary (Stoic) approach to emotions in Western thought.[23]

Emotional beliefs are problematic, for the Stoics, if they ascribe value to some person or thing outside of the individual's control. On the Stoic account, the only thing that has value is one's own good character. Furthermore, having a virtuous character is not dependent on external circumstances. Hence, if emotions are judgments that something outside of our control has value, they are false beliefs. The conviction that emotions are judgments is necessary for their prescription of apathy (living without *pathos*). If emotions are not judgments, it is not clear how we can be said to have control over them and how we might go about completely ridding ourselves of them. Also, as a retort to Aristotle, they add that if emotions are not judgments, it is not clear that we could have any kind of rational control over them, and yet we clearly do. The Stoic student can come to realize that his material possessions do not reflect his true value. His emotions about those possessions will thereby change, and perhaps even change rather quickly.

Most contemporary cognitive theorists of emotion follow the Stoic line that beliefs are necessary and sufficient for emotions while affects are neither necessary nor sufficient. Yet, these cognitive theorists do not agree with the Stoics that the emotions are necessarily mistaken judgments; they believe that they can inherit a Stoic account that is more friendly toward the emotions.

In Martha Nussbaum's Aristotelian and Stoic account, for example, emotions are value judgments. Nussbaum uses the titular metaphor from her book *Upheavals of Thought* to describe emotions, holding that they are typically conscious mental preoccupations. She does not mean that we must constantly attend to an emotion; rather she means that it has cognitive content. She identifies emotions with eudaimonistic judgments:

> So we appear to have type-identities between emotions and judgments—or, to put the matter more elastically, looking ahead, between emotions and value laden cognitive states. Emotions can be defined in terms of these value laden recognitions alone, although we must recognize that some feelings of tumult or "arousal" will often accompany them, and sometimes [affective] feelings of a more type-specific kind, and although we must recall that they are at every point embodied. If we want to add this very general stipulation to the definition, we may do so though the proviso that we are talking about only the likely case, in order to retain the possibility of recognizing nonconscious emotions.[24]

Nussbaum agrees that emotions often involve affective states, but she rejects the idea that affective states should be given definitional priority. In addition to her example of the unconscious fear of death, which she holds to be an emotion, she gives the example of anger at/after her mother's death, an emotion that lasted for days and sometimes went without a physical manifestation, yet at other times expressed itself through physical symptoms.

We may not often experience the cognitive content without the bodily feeling of an emotion, but cognitive theories capture an important insight: we would not deny that someone is sad in the absence of certain privileged bodily feelings—simply saying "I am sad" would presumably be enough—but we might deny that a person is sad in the absence of a reason or occasion for sadness. In such a case we would look harder for a reason, or we would deny that the person is really sad... perhaps he is just tired? In this sense, depression, if it is taken to mean sadness for no reason, is thought to be an emotional disorder.

For Robert Solomon, another contemporary follower of the Stoic (turned Sartrian) philosophy of emotion, "emotions are judgments" ("normative and often moral judgments"),[25] but this does not mean that emotions are propositional attitudes; rather, "emotions are subjective engagements in the world."[26] A judgment is an "engagement" because both are intentional.[27] Solomon focuses on the notion of emotional engagement to suggest that emotions can be willful or are something we can be "caught up in." Trying to include their affective aspect, Solomon calls emotions "judgments of the body."[28] He notes that physical affects often accompany these judgments, but he argues that emotions should not be identified with affects because affects are not intentional—they are not *about* something—and emotions are. As we can see, contemporary cognitive theorists have moved to include affect in their theories, but they hold that emotions are better understood in terms of their cognitive content.

For Nussbaum, the upshot is that emotions teach us about ourselves and our values. She believes that we often rationally or unconsciously ignore our relational attachments, and we ought not because they are our moral attachments. For Solomon, following Sartre, the moral implication of cognitivism is that we are always responsible for our emotions. Sartre posits that even the most seemingly involuntary gestures are the products of disassociated wishes and should therefore be subject to ethical scrutiny as though they were voluntary. Being a bit more practical, Solomon focuses on the way that emotional displays often serve unconscious purposes and should therefore be taken to task. He gives us the example of a husband who picks a fight with his wife, and ends up feeling very angry; this emotion has the effect of ("magically," Sartre would say) getting him out of going to a party to which he did not want to go in the first place. There seems to be something blameworthy about this dishonesty and the selfishness that motivates it, and the husband is either unaware of this moral dimension to his emotion or he is denying it.[29]

To sum up: cognitive theories of emotion hold that emotions, like cognitions, are always about something. Emotions involve bodily feelings, but those feelings do not themselves tell us about the emotion.

MIND AND BODY

Cognitive and affective positions are unfortunately all too often caricatured. Cognitive theorists have said that affective approaches hold that emotions are "dumb" or "unthinking movements" because they do not involve the mind at all; affective theorists have said that cognitive theorists think that emotions are "disembodied" or that they are fake, intentionally caused displays.[30] Neither characterization is true. Both approaches have some explanatory power. In fact, as we can begin to see, there are important ways in which they agree. One might be tempted to say that affective and cognitive theorists disagree because the former identify emotions with the body and the latter identify emotions with the mind, but, as we have seen, the debate is somewhat more complicated than that. Emotions are interesting precisely because they reside at the intersection of the mental and physical, and each side of the debate acknowledges this in its own way.

In order to fully address the differences between these two positions, and hopefully gesture toward a reconciliation, we must first consider whether or not there are in fact substantial differences. Possible differences include whether or not the majority of emotions are likely to be caused by mental events or external events, the nature of perception and its relationship to judgment, the extent to which we have control over our emotions, and whether or not moods should count in our inquiry. Addressing these topics will provide a sense of the precise way in which emotions reside at the intersection between mind and body. Additionally, charting a more subtle understanding of emotion will begin to show us the way we can avoid the three caricatured profiles of emotionality outlined in the previous chapter and what a good emotional comportment might be.

We can see the extent of the overlap between cognitive and affective theories, as well as their essential disagreements, if we examine the way that Kant and Descartes define emotion. Descartes and Kant both provide a similar definition of emotion (and both are pretty good). They provide a good starting point for any discussion of emotion by first distinguishing between emotions and abstract thoughts, on one hand, and emotions and purely physical sensations, like hunger or pain, on the other hand. In section 27 of *Passions of the Soul*, Descartes lays out a set of distinctions. First, he distinguishes between "actions" and "passions." Roughly speaking, in Descartes's terminology, actions are things that we do; passions are things that happen to us. We can call passions "experiences." Of our experiences (second distinction), we experience our own actions and we experience things that are going on in our bodies. Of the things that are going on in our bodies (third distinction), some are more mental, like imaginings, and some are more physical. Of our physical experiences (fourth distinction), some of them refer to an external object, some refer to the body, and some, Descartes says, refer to the soul, or the mind itself. Given this theoretical apparatus, emotions are those physical experiences that refer to, or are about, the mind (passions *of* the soul).

Emotions, although they involve bodily feelings, do not tell us about the state of the body; they tell us about the state of the self. So, for example, following Descartes's example, physical pain is a product of physical injury or stress in the body; it tells us something about what is going on in the body. Any of the five senses, to give another example, sight or taste or touch, tell us about an external object. Emotions, although they are physiological experiences, are different still. Anger, for example, tells us something about ourselves. It tells us that our plans are being frustrated. They both may have the same material cause; the neurocircuitry might be the same for both physical pain and emotional pain. Their efficient cause might even be the same: if someone kicks me it would give me both physical pain and make me angry. Nevertheless, they have a different formal cause; they are "about" something different. Emotional affects necessarily have something to do with the self.

Kant further spells out the way that "passions" can be *of* the soul. Kant agrees that emotions (cognitively caused feelings) are neither purely physical sensations nor abstract thoughts.[31] Feelings (of pleasure and pain) can be the effect of all of the mind's faculties (sensibility, imagination, understanding, and reason; A 230).[32] Kant frequently discusses feelings, like awe or respect, that are caused by abstract thoughts. For him, bodily feelings are not "about" certain things but the feelings *of* certain thoughts (DR 211).[33] If Kant is right about this, he can help us advance the cognitive/affective debate.[34] Certain thoughts cause bodily feelings. (This fact strikes me as both obvious and very interesting.) Kant probably does not think that all thoughts cause an effect in the body, but that is an open question. Unlike Descartes, Kant does not see the feeling as a reflex of the body but as an effect of thought, which may or may not be freely undertaken. We need not say that it was the kick itself that made me mad, but instead we can say that it was my perception or thinking of it as a slight or obstacle that caused the angry feelings to arise physically.

Following Descartes and Kant, we should distinguish emotions from physical pain and abstract thoughts.[35] Physical pain, itchiness that comes from a bug bite, the burn of a skinned knee, hunger or thirst, having to go to the bathroom—none of these seem like emotions. Then there is the strange fact that thoughts and perceptions can also cause physical feelings. Even if these feelings are very general states of arousal or depression, there is no doubt that our thoughts make us have feelings. Similarly, these feelings can sometimes spawn further thoughts, which themselves create feelings. We will call the complex, temporal relationships between physiological sensations and thinking *mood*, and here we can begin to see that understanding mood is a necessary part of understanding emotion and, thereby, developing self-understanding.

On the other hand, we need to distinguish between emotions and purely abstract thoughts. There is no doubt that even the most abstract thinking can *cause* emotions. Abstract theoretical physics, for example, can give us a feeling of beauty or wonderment, even joy or divinity (but perhaps that is only if we grasp its connection to us). We must make a distinction between thoughts that are themselves emotional (they themselves cause feelings) and thoughts that may be contingently related to emotional reactions (thoughts about those thoughts cause feelings). For example, the evaluation that a certain theorem seems to be totally unimportant makes me feel bored. The theorem itself, or my comprehension of it, is not what makes me feel bored. We can feel anxious about finding the right answer, fearful of our own inabilities, or delight at completing a challenge, but these emotions are about our relationship to the abstract thoughts, not a product of the abstract content itself. If we are going to argue that some thinking is very "lively," to use Hume's term for present experience as opposed to abstract thought, and causes a lot of physical feeling, then it only stands to reason that some thinking is not very lively, and perhaps almost dead. Feelings, for Kant, are physiological events necessarily involving some form of pleasure or pain; they are caused by mental events, but they are not themselves mental events.

Affective and cognitive theorists ought to agree on this much: we can make an important and meaningful distinction between emotions and abstract thoughts, on one hand, and emotions and merely physical sensations, on the other. Still, although it *seems* wrong both to identify emotions purely with thoughts and to identify emotions with other types of physical feelings, like pain or hunger, it might be impossible to insist on definitional homogeneity even about this seemingly non-controversial suggestion. Some affective theorists believe that there is a meaningful relationship between purely physiological responses, like hunger or pain, and emotions, while some cognitive theorists, like Nussbaum, believe that some thoughts without a particular physiological component (or even a conscious mental component, for that matter) can be emotions.

In his textbook on emotion, *Emotion Explained*, Edmund Rolls devotes chapters to hunger, thirst, drug addiction, and sexual desire (as he assumes that sexual desire is a purely biological drive).[36] These biological drives that Rolls includes with emotions are not usually given that classification in common discourse—as might be the case with startle. Startle is physiologically identical to fear, but is not, merely for that reason, the same as fear. Jenefer Robinson has defended the opposite position.[37] She criticizes what she calls the "judgmentalist" philosophical theories of emotion for ignoring the "primitive side" of emotion.[38] Following Ekman and the Darwinian tradition, she argues that startle should be considered an emotion because it involves a characteristic pattern of neural firing and a characteristic facial expression. She argues that it is a "developmentally early form" of fear and surprise. Although Robinson uses LeDoux to challenge the idea that emotions require conscious thought, she argues that the startle response is an implicit judgment and that, because of this, it should be taken as the prototypical emotion:

> Emotional response should be thought of, on the model of the startle response, as a response that focuses our attention on (makes salient) and registers as significant to the goals (wants, motives) of the organism, something in the perceived (remembered, imagined environment); this response characteristically consists in motor and autonomic nervous system change.[39]

This very general characterization does show the similarities between startle and emotion, but there are also important differences. For example, we are startled well before we know whether or not the event is relevant to our desires, and for that reason the startle response is often mistaken. Ekman, on the other hand, holds that startle is not an emotion because it cannot be inhibited or simulated and because it is reliably caused by a loud noise, while emotions are not reliably caused by any one general thing.[40] We might say the same thing about fear of snakes or spiders, insofar as that fear is innate.

While I agree with Ekman that startle is not an emotion, it seems that people might use the term "emotion" in different ways. If we are going to say that hunger, startle, or the life-long fear of death are emotions, then we must give up on "emotion" being a natural kind term.[41] Insisting that examples that fit our definition must be counted, while insisting that examples that do not fit our definition must be excluded, does nothing but beg the question. Similarly, the fact that other emotions involve some kind of reflexive responsiveness, physiological processes, or thoughts, does not count as sufficient reason to use any of these emotional experiences as *prototypical* emotions that can thereby be used transitively to inform us about other emotions. The fact that LeDoux has supposedly found a neural pathway for startle that entirely bypasses the neocortex is quite remarkable, but it would be unscientific to thereby deduce that other emotions also involve neural pathways that bypass the conscious thought of the neocortex.[42]

The lesson here is that even if we can agree on a very basic definition of emotion (I suggested excluding abstract thought and physical pleasure and pain), we still must be careful when we rely on facts about one emotion to make conclusions about other emotions. Emotions are very different from each other, and there is no reason to ignore or suppress these differences. We can call this the rule of expecting difference: if we expect that each emotion will be significantly different than the last, we will be prepared to put in the effort necessary to understand and evaluate them.

INTERNAL OR EXTERNAL?

One important difference between affective and cognitive theorists seems to be their accounts of the causes of emotions and the extent of control we have over these causes. The affective approach seems to privilege the explanatory power of external stimuli in the life of the emotion, while the cognitive approach seems to privilege the mental life or mental causes of the emotion.

The first thing we should note is that there is not as clear and easy a distinction between an internal and an external cause (or between judgment and perception) as affective theorists would have it. We can often only see the role that judgments play in emotions when those judgments are false. If I get jumpy and excited when I see a car pull into my driveway, we would say that I am reacting to the fact (the perception) that my date is here. If that car turns around and my date subsequently arrives on a bicycle, we would say that my original reaction was a response to my belief that my date would be driving.

Nevertheless, it does not make sense to call the cause of the emotion a perception at one time and a belief at the next.[43]

Many who adopt affective or perceptual theories of emotion—indeed many people overall—would assume that perception and conception are two different mental processes. If James's theory of emotion is to have critical leverage against a cognitive theory of emotion, then perception must refer to some mental process that is more immediate than conception (just as LeDoux has found a neural pathway that bypasses the neocortex). As it turns out, James's distinction between perception and conception will not help the affective theorist. His use of the term "perception" refers to immediate experience, including a stream of consciousness experiences—the sort of thinking that includes the kind of mental talking to oneself that is based in concepts and believed to reside in the neocortex. His use of the term "concept," on the other hand, is more akin to what we would normally call "theorizing": more abstract, philosophical thoughts. If we cannot draw a hard line between perception and concepts, we must question the ability of affective theories to differentiate perception from understanding and therefore affective theories themselves from cognitive theories.

Prinz agrees that emotions involve appraisals, but he aligns these appraisals with perception rather than judgment. Both James and Prinz take perception as the model for emotion, but we cannot conclude thereby that they take emotions to be any less conceptual than cognitive theorists do. James would likely protest that emotions do not involve concepts, but this is due to his derogatory view of concepts, not of emotions. Over the course of his life, James became more skeptical of the legitimacy of abstract conceptual thought and more convinced that truth was conveyed immediately through practice.[44] In his *Varieties of Religious Experience*, James's discussion of truth borders on mysticism. He describes spontaneous religious conversion as the achievement of a more harmonious integration of beliefs and feelings. In other words, conversion is the most important, although just one, example of emotion playing the lead in showing conscious (conceptual) thought the truth. For James, concepts refer to more abstract mental processes, not instruments that facilitate normal experience. James follows Peirce in understanding the beliefs that inform normal decision-making processes as mental habits; in "The Sentiment of Rationality," James comes close to arguing that rationality itself is merely a sentiment. Toward the end of his life he became more suspicious of conceptual thought in general, suggesting in his "A Pluralistic Universe" that it has a tendency to distort reality.[45]

James's theory of perception is based in his neutral monism, which holds that the "mind" and "body" are abstractions that refer to one underlying substance that is itself neither mental nor physical. Similarly, his descriptions of experience combine what we would normally take to be both mental and physical aspects.[46] Perception, for James, is nearly indistinguishable from conscious recognition because perception itself already represents a cognitively organized form of sensation.[47] Emotions can only be so closely tied to perceptions because perceptions are not of "the given," as Quine argues that empiricists would have it, but are already laden with content.[48] It is not just that we see a box outside of the door, we see the box *as* a gift that we have been happily anticipating. It is the latter *conceptual* meaning, not the box itself—its pure physicality of shape and color—that is a necessary component of an emotion. Using the term "conceptual" instead of "cognitive" represents progress in overcoming the debate because cognitive theorists are not committed to the idea that the content of emotions has already been explicitly cognized before the emotion occurs, merely the idea that it is there and can be made explicit, even if only after the fact. In the example of the gift, the person most likely does not explicitly think, "There is the gift from my mom that will probably contain my grandmother's necklace that I used to love trying on when I was a little girl," and yet all of those thoughts (and more) are in some sense present in the perception.

Just as Kant takes his model for experience from judgment, James's account of perception includes what we would normally refer to as conceptual recognition. When we look at James's philosophy more holistically, it becomes difficult to use it as a wedge between biological and psychological approaches to emotion. If we follow Solomon in asserting that judgments need not be propositional attitudes, but are rather ways of representing objects, then we can argue that James's notion of perception similarly offers a theory of "engagement" with world and that this engagement is mental.

It is easy to agree that different emotions involve different "core relational themes"; this characterization of each emotion seems to involve little more than telling us what the names for each emotion mean. For example, fright expresses the fact that one believes that one is facing some kind of immediate danger.[49] Emotions, for Prinz, "represent" core relational themes, but core relational themes do not count as "judgments."[50] It is unclear whether or not it is appropriate to use the term "judgment" to refer to the kind of thinking that core relational themes imply. Perhaps "conceptualization" makes more sense. These descriptions are conceptual, but explicit conceptualization may or may not have come before physiological response.[51] The majority of theorists seem to agree that emotions can occur rather quickly, as if they were not connected to conscious thoughts; perhaps, in these cases, we might say that emotions involve snap judgments.

It seems that a cognitive theorist should be willing to accept a robust revision of James's theory of perception in place of "judgment" or "belief." In truth, in calling emotions "embodied appraisals," Prinz sounds a lot like Solomon or Nussbaum.[52] Prinz's argument shows, perhaps inadvertently, the way that the affective position is conceptual on its own terms. The affective account relies on the complex meaning of the environmental or internal cue to cause the affect.[53] In response to the possibility that affects precede judgments, Solomon retorts, "there is still cognition there," referring not to the presence of slow and deliberate judgments, but to the possibility of giving a rational account of the situation.[54]

On Kant's account, it would not make sense to say that either the feelings or the mental content has temporal or logical priority. Just as the pleasure of the sweetness of a peach is a part of tasting it, the pleasurable feeling of the free play of the cognitive powers is a part of experiencing beauty. Kant's account further helps us to adjudicate the debate about the meaning of perception because bodily feelings, for Kant, can be caused by cognition at any level (perception, imagination, understanding, and reason). Furthermore, perception for Kant, is a part of the cognitive faculty, albeit an inchoate one.

While it is the case that some events are caused by memories or our calling certain ideas to mind, while other emotions are involved more with present occurrences (and we should not privilege either type as being more authentic), what is going on here is that the very same emotion can be described in two different ways, either as being caused by an external event or as being caused by the subject herself. For an example, with Prinz's notion of "calibration files," why should we say that an *event* triggers certain calibration files and causes an emotion and not instead say that the person herself (with her personal history, upbringing, culture, and ideas) causes the particular emotion at the occasion of certain events?

CONTROLLING EMOTION

Calling the cause of emotion a perception (or calling the emotion itself a perception) makes it seem like the affect is a direct effect of external stimuli; calling the emotion a judgment makes it seem like emotion is dependent on the subject's beliefs.[55] The relative internality of the cause of the emotion reveals itself to be a sticking point only because it suggests a certain degree of control over the emotion. If we ourselves are the causes of our emotions—our concepts, our ideas, our past experiences—then it seems that we are in control of our emotions. If outside events cause the emotion, then the emotion seems to be outside of our control.

While it is no doubt the case that some emotions seem to follow automatically from certain perceptions, a deeper consideration of the nature of perception reveals that these emotional pathways cannot be entirely biologically innate or behaviorally conditioned: they are simply too variable. They must be open and flexible to accommodate new experiences. The number of different objects and personal experiences that could signify personal loss or the frustration of plans, and hence trigger sadness or anger, are so vast that positing a direct neural pathway between all of them and different emotional centers in the brain is absurd. Similarly, it would seem to suggest that there are too many new experiences of sadness or anger that we would miss just because they did not sufficiently resemble our past triggers of anger or sadness. The brain is a much more sophisticated piece of adaptive machinery than the theory of psychological mechanisms suggests. Furthermore, no object or experience reliably causes emotions, not even within the same person. Seeing a trophy today might make me feel excited, while tomorrow it might make me feel proud, or perhaps angry. The only way to explain this emotional elasticity is to say that the trophy offers a different conceptual significance at different times.

Prinz denies that emotions are cognitive because he takes cognition to involve an *act* that is within the subject's *control*. Retreading some Aristotelian ground, he briefly admits that there is a sense in which emotions are under our control and a sense in which they are not under our control:

> Emotions are voluntary in a double sense. Thinking about something in the right way can certainly influence our emotions, and calibration files can be modified through education and experience. We exert control over emotions by choosing what to think about, and by cultivating calibration files. But emotions are also involuntary in a double sense. First, the thoughts and images contained in an established calibration file may set off emotions automatically. If one happens, by choice, to activate a representation in a calibration file, an emotion will ensue. Second, once an emotion has been initiated, we cannot alter it by direct intervention. Initiation pathways and response pathways both operate without the luxury of control.[56]

We should take note here and similarly make these distinctions about the different ways we do and do not have control over our emotions. When we say that emotions are the feelings that come from certain thoughts, we must note the different ways in which we have control over thoughts and over feelings. On one hand, although we seldom go about trying to change our thoughts, and it is not immediately obvious how we should go about doing it, we can supposedly control the thoughts that we have. Altering thoughts can alter feelings. Prinz also points out the importance of behavioral self-training. A fear of public speaking, for example, will respond better to experiences of successful public speaking than it will to internal coaching. We cannot simply tell ourselves anything we want to (and believe it); therefore, there is good reason to point out that our thoughts are beliefs based on facts, and they are only under our control to a limited extent. On the other hand, we can alter feelings by doing things that create or inhibit them directly: we can take a nap, a bath, or medicine, for example. Prinz is right in that emotions can sometimes strike us quickly and all of a sudden. There seems to be little we can do in the seconds in which the feeling is developing to stop it dead in its tracks. (Although, there is often little reason why we would want to.) Nevertheless, Prinz is wrong to say that "once an emotion has been initiated, we cannot alter it with direct intervention." There are plenty of examples of people trying to control their emotions, and perhaps sometimes succeeding. Although being startled might trigger a rush of adrenaline, for example, we can decide if we want to "cool off" or stay excited, and we do things and think thoughts to help in either process.

If a feeling creates a desire to behave in a certain way, we have a choice about whether or not we will carry out that behavior. In his *Religion Within the Bounds of Mere Reason*, Kant offers us the theory that we always rationally assent to our desires before we enact them.[57] It is handy to say that we were overpowered by an emotion, but, according to Kant, we actually always choose whether or not we will respond to a particular feeling in a certain way, no matter how strong it is.

Similarly, we can alter our physical feelings both in the short and long term, mentally, physiologically, and behaviorally. The success of these attempts at change are dependent on the type of emotion and facts about the situation. Nevertheless, Prinz is right to suggest that the *connection* between certain cognitive events and the pleasurable or painful effects they cause (feelings) is not within the sphere of direct cognitive control. *That* certain thoughts cause certain feelings is often a biological fact about us. In addition, we cannot simply say to the heart, "Stop beating so fast," and make it so. In this sense, again, Prinz is right to say that we cannot stop an emotion "dead in its tracks"—but we might be able to stop it slowly by taking a couple of deep breaths. We should also note, from experience, that one thing that seems to calm our feelings is coming to cognitive insight about their causes. If our feelings often prompt bad habits, either unhealthy or immoral, then it is our responsibility to figure out ways to try to change them. While the most effective ways of doing this are perhaps relative to the situation and even perhaps not yet discovered, these *methods* are also susceptible to moral evaluation: we must make sure that we similarly preserve respect for our rational nature.

The Stoics held that we can change our emotions by changing our judgments; they did not realize that we are not radically free to change our judgments. Cognition is an evolutionary adaptation, and it is not radically open to variation in function. We cannot change the facts, and we also cannot change some of our basic values, like self-preservation and the basic physical and psychological needs that accompany it. Nor do we often experience ourselves as directing cognition; it is rather the case that the topic or experiences at hand directs cognition.[58] We can meditate or write in a journal, but we cannot will away our "attachments" and we cannot learn or solve problems without the help of external knowledge and models.

Along these lines, it can sometimes feel like feelings are in control. Some feelings, for example, seem to be much bigger than thoughts. I was overcome by sadness following my father's death, and yet the thoughts that accompanied his death were relatively few and simple: "My dad is dead." "My daughter will never meet her grandfather." These thoughts are so simple, and pace Nussbaum or the Stoics, I did not always attach much value to these thoughts. I did not think, "My dad is dead and that is a bad thing," yet the value seemed to speak for itself in the emotion, and it spoke much more loudly and powerfully than the simple, cold fact of his death. When the feelings were gone, there was nothing left.

MOOD

Surprisingly, it turns out that the most important difference between affective and cognitive approaches centers on the question of whether or not a theory of emotion should include moods. We cannot turn to evidence to answer this question; the answer is merely a presumption of study. A nonarbitrary or nonquestion-begging answer is not possible. (An analogy: Is drinking a kind of eating? If I am going to study eating, do I have to study drinking too? If not, can I make a sharp distinction between them? What about soup?) Although this looks like a question about the fact of the matter—one that we should be able to answer by consulting the definition of emotion—in fact the different definitions have been formulated based on an assumption about the answer. We can only answer the question pragmatically; in other words, the answer depends on what you hope to get out of your definition of emotion. If you want to make scientific observations about the relationship between certain environmental causes and biological effects, then including mood in your study would be too cumbersome. If you want to better understand the causes of behavior (your own included) in order to try to become a better person, moods are impossible to ignore.[59]

There are bad moods and there are good moods. With both of these phrases we seem to be referring to some kind of affective phenomenon. Someone in a bad mood will frown, be quarrelsome, withdrawn, self-focused; someone in a good mood will be smiley, quick-paced, and bubbly. The causes of these moods could be physiological: not enough sleep, pain, or drug consumption (like caffeine). The causes of these moods might also be cognitive: worry about one's health, stress about money, concern about relationships. Although many will say that moods are like emotions but less intense and longer lasting, there does not seem to be a clear dividing line between them. If emotions turn into moods, there is no nonarbitrary time point at which this change happens. When moods elicit emotions, we are tempted to reduce the emotion to the mood. The lesson here is that the context of an emotion is of the utmost importance—mood helps us to characterize the context—and that an understanding of context, far from challenging the universality of emotion, is a precondition for self-understanding, self-evaluation, and sympathy.

Prinz's approach is predicated on a distinction between "state emotions" and "attitudinal emotions," the latter involving a more complex relationship between thoughts and affects over time.[60] This distinction suggests that moods should not count as emotions. Nussbaum, on the other hand, identifies an emotion, the fear of death, that lasts a lifetime.[61] Most theorists of mood describe it as an affective state, and so it seems that although affective theorists are the ones more likely to attempt to exclude mood from consideration, their theory gives them the least warrant for doing so.

Nevertheless, both affective and cognitive theories of emotion would most likely agree that specific thought-affect combinations will get called "emotions" as opposed to "thoughts" or "moods" because they fit certain nominally typical response patterns. Yet these prototypical response patterns, or basic emotions, are simplifications. Real emotions, which are often inextricably linked to moods, are often much more complicated than any account of a single emotional response suggests.[62]

Recently I met with a student whose mother had just died. Her mother was diagnosed with a terminal disease in the middle of the semester, and she died in the last week. After the student came back from her trip home to go to the funeral, I met with her to talk about her final paper and the possibility of taking an incomplete in the class. I asked her why she did not just take the semester off (when her mother was diagnosed), and she answered "I wish," indicating that she did not take that to be possible. Although she made every effort to focus on talking with me about the final paper, it was still evident that she had suffered a great loss. She moved slowly and did not smile much. When she did smile, there was an immediate look of discomfort and confusion that followed. Her eye contact was more slow and steady, until the topic veered toward anything too positive or negative and then her focus was forcibly cut off. Someone in this situation is caught in a bind: she does feel sad, but she cannot feel sad. The fact that she does feel sad means that it is not only physically difficult but morally undesirable for her to focus on anything but her mother's death, yet the fact that it would be financially unwise for her to take the semester off and risky for her to take an incomplete means that she cannot devote her time to crying and the other appropriate expressions of sadness.

Are emotions coterminous with their expression? We often cover up our emotions. Ironically enough (if taken in the context of those who would identify emotion with its expression, like crying), Kant points out that real emotional expression is almost always stifled and awkward. Only in fake emotional expression do we see the emotion expressed perfectly in its typical way, like crying or yelling. Kant takes this to be evidence that emotions cannot be reduced to their affects.[63] A focus on physiological affect, as it is construed in a short-term, visible way, would miss this important fact, while a focus on mood would address it since a discussion of mood invokes the complexities of the situation.

Although we normally do not use the word "mood" for the above case, this experience of grief could either be called an emotion or a mood. It is a mood because it describes the affective quality of lived consciousness over time. What does this example tell us about moods?

They, like emotions, are difficult to make sense of without a discussion of their context. That context can sometimes be best described in biological terms and sometimes in terms of events and their social and moral meanings. Excluding mood from a theory of emotion does not seem to be based on a reflection about our experiences of moods and emotions and the relationship between them; instead it seems to be based on a dogmatic desire to privilege some kinds of emotional experiences over others in order to preserve a simple definition of the term.

Second, the above example reminds us that most human emotions are complicated. An understanding of context is essential for an understanding of emotion because emotions blend into and accompany other emotions so easily. The notion of mood usually captures this truth. When I spill my hot soup down the side of the public microwave in the library and even over the bins of bagged plastic utensils, I feel startled, annoyed at myself, sad, perhaps even defeated. I probably even feel other emotions that I repress, such as embarrassment. If I were a more irascible person, I would feel angry. Hopefully I can meet the situation with humor: "What a day!" Perhaps I can even think of a punch line, like "Now my soup is low-cal!"[64] Different emotions may surface as the situation passes and it is more safe to acknowledge them. Only later might I fully acknowledge the true force of the embarrassment. On the other hand, components of the experience might go underground, and the affects might remain in the form of mood or might blend with other moods, which will have various other causes, both physical and mental. Even still, if I am in an exceptionally bad mood to begin with, this minor event might have explosive consequences.

Without an examination of mood we cannot accurately understand the relative causal roles that affect and cognition play. It is disingenuous for affective theorists to exclude mood (or attitudinal emotions) and then insist that the proximal cause of the emotion is a relatively simple external stimulus. Even if affective theorists insist that the physical affect is the essence of the emotion, we will have to take into consideration the *previous* affective state of the organism in order to explain any subsequent affective response. After all, living organisms are not like computers, forever freshly responding to new inputs with the appropriate outputs. Unlike computers, we get tired, cranky, defeated, depressed—we have moods.

We must be careful to analyze our moods, parsing out emotional and merely physical causes. We sometimes mistake genuine emotional responses for mere physical symptoms, like having a headache or a stomach ache, and vice versa. To an extreme degree, failure to recognize or to correctly label emotion is called "alexithymia," which is caused by, for whatever reason, a lack of access to the cognitive dimension of emotions. Although it is decidedly *not* the case that all illness is the result of negative affects and emotions, as some "New Age" thinkers suggest, emotional stress is a major cause of many diseases and unhealthy life choices, and we must have some degree of psychological acumen in order to stay physically healthy. On the other hand, we sometimes mistake merely physical responses, like hunger or tiredness, as bona fide emotional responses. We need to know when it is better to ignore our emotions and simply wait for them to pass. Similarly, manipulating and stimulating physical affects, for example, by means of laughing or exercise, can be a truly healthy way of freeing oneself from a mildly bad mood or a small emotional setback. Most people have a working understanding of mood and mood alteration, and these techniques are a part of the basic path we all follow toward physical health and psychological well-being.

Even if we agree that there are basic emotions with universal, characteristic expressions, as I am inclined to, there is no reason thereby to discount the importance of mood.[65] It would be absurd to insist that moods are *less* emotional because they are not as easily characterized as the *basic* emotions—this would confuse a difference in kind with a difference of degree. We thematize basic emotions and emotion concepts in order to help us make sense of complicated emotions and our lived emotional experiences, not the other way around.

A consideration of mood also prompts us to ask whether or not all experience is emotional. Damasio's Spinozistic theory of consciousness seems to suggest that *all* experience has an affective, or bodily, component.[66] This monism is also a Heideggerian idea—the idea that Dasein (or human consciousness) is *always* in a mood—and, as we have seen, it is at home in James's proto-phenomenological philosophy, as with the idea that rationality is itself merely a sentiment or mood.[67] (Hume's notion that experience is distinguishable from thought by—*the feeling of*—its liveliness presages James and suggests a way in which empiricism might lead necessarily to an affect-based phenomenology.)[68]

Paul Ekman, through cross-cultural studies, determined that there are universal characteristic facial expressions for six basic emotions: anger, disgust, fear, shame, joy, sadness, and surprise. In subsequent studies came evidence that contempt also has a universal, characteristic facial expression.[69] Even when we rule out physical reflexes, physical pain and pleasure, and purely abstract thoughts—which I am happy to do—it still seems that there are a lot of feelings that may or may not be emotions: humor, boredom, interest, uncertainty, curiosity... These are more intellectual feelings, and yet they all seem to have (at least mild) affects.

When we begin down Ekman's path of looking at characteristic facial expression, we become amazed at how much emotional information we get from one's face and gestures. Ekman himself began his career in noting the difficulty in measuring nonverbal communication.[70] Nothing stops us from saying that all unintentional nonverbal communication is emotional communication.[71] A compelling exploration of gesture and nonverbal expression can be found in Esther Shalev-Gert's "Between Telling and Listening."[72] Shalev-Gert offers video clips of people's expressions while they are listening (to what we do not know) or while they are saying something (what, again, we know not). It is amazing how much the audience can glean about these people's emotional state. Of course, the audience might be wrong, but what is even more amazing is that the sympathetic response of the viewer is heightened—the emotional state of the video subject seems to be contagious—with the context removed. Human beings are naturally very good at communicating nonverbally; we, of course, might incorrectly interpret a bodily or facial expression, but that possibility seems biologically unlikely for the majority of our interpretations since this hypo-linguistic ability is necessary for living in groups—is this not the way animals communicate?—as well as a precondition for the learning of language by infants and toddlers.[73]

It is also possible to argue that, not just present experience, but also memories are affective. Carruthers, in her study of medieval rhetoric, argues that

> Some traditions in ancient philosophy also recognized an emotional component in all memory. Memory images are composed of two elements: a "likeness" (*similitudo*) that serves as a cognitive cue or token to the "matter" or *res* being remembered, and *intentio* or the "inclination" or "attitude" we have to the remembered experience, which helps both to classify and retrieve it. Thus, memories are all images, and they are all and always emotionally "colored."[74]

Entertaining Heidegger's insight that we are always in a mood reveals a further insight: that emotion is a quality of temporality. Mood is precisely the elongation of consciousness through time.[75] We saw in the last chapter that it is now popular to insist that we should live our lives "in the moment." Unfortunately, that injunction is exactly the call to live life without emotion. Emotions impinge on the present. They involve memories or anticipations. They connect us to the past and to the future. Emotions make demands on our present consciousness as the voice of memories, values, and commitments. We should not try to live life "in the moment"; remembering the past and planning the future is what makes us human. Instead we need to make sure that we are doing justice to the past by making the best plans for the future.

EMOTIONS ARE EVERYWHERE

Although we have only surveyed cognitive and affective theories, we might also remark that there can be as many different types of theories of emotion as there are seen to be parts or faculties of the mind. Emotions are everywhere: they relate to everything that we do and every part of the brain with which we do it. So, for example, we can also say that there are conative theories of emotion, that is, theories that hold that emotions are desires (or states of action readiness).[76] Prinz, whose Humean project parallels my Kantian one, makes a similar point, writing: "Every obvious part of a typical emotion episode is identified as the essential part of an emotion by some theory."[77] For example, emotions are related to thinking, so there are cognitive theories of emotion; they are related to desires, so there are conative theories of emotion, and so on. Prinz takes the fact that emotions involve so many different psychological parts to be a problem; he calls it "the Problem of Parts," and he attempts to solve this "problem" by identifying a part that is necessary and sufficient for the experience of emotion, thereby formulating an essential definition of emotion.[78] This method will only be successful if it truly does account for all emotions and does not simply beg the question about what counts as an emotion and what does not.

Prinz believes that he solves the "Problem of Parts," but instead he merely creates the problem of marginalized parts, as he ignores the importance of attitudinal emotions (moods), insisting that concepts are not internal to the experience of emotion. I urge us to instead begin more modestly, with observation, so that we can appreciate and understand the vast and varied world of emotional phenomena.

Dogmatic adherence to either cognitive or affective theory misses more subtle differences between different emotions, as I have shown here, and can also be understood as discipline-specific assumptions. Affective approaches tend to take a scientific, third-person perspective, while cognitive approaches begin with an introspective, therapeutic approach. As long as we keep in mind that emotions involve both the mind and the body, and that the majority of human emotions are rather complex, then it seems that affective and cognitive theorists can each go about their business without offending each other. (Of course, when the biological crosses over into the therapeutic, the danger that people might misunderstand their own emotions results. See the "Conclusion" for a further discussion of the need for closer attention to disciplinary boundaries in the study of emotion.)

Calhoun objects to the cognitive approach because, as she argues, emotions and beliefs are "logically and ontologically" distinct categories.[79] Calhoun is almost right. Thoughts and affects seem to be logically and ontologically distinct, but emotions are exactly those experiences that blend them.[80] Emotions are interesting precisely because, although we commonly assume a dichotomy between physical and mental experience, even an unsophisticated account of an emotion belies it. Most emotional affects are obviously psychosomatic: we commonly recognize that we are able to "worry ourselves sick" or "get worked up." In addition, in the hum of everyday experience, we have a multitude of physiological feelings, and it takes work for us to figure out if a feeling is mentally or physically wrought.

Kant offers us a satisfying resolution to this debate with his notion that feelings arise from all of the functions of the cognitive faculty. He is steadfast in his insistence that feeling and cognition are two different things, but thoughts and perceptions (which he, with other sensations, classes as a part of the lower cognitive faculty) *feel* a certain way. We need not even say that the cognitions come first since there are many cases in which we are more preoccupied by the pleasure or pain than the thoughts; nevertheless, they go hand in hand.

For Kant, the first rule of developing virtue is "know thyself." Before we can judge whether or not our actions are morally acceptable or even morally praiseworthy, we must know what our motivations are. That sounds simple, but it is not. Having a good working understanding of emotion, which includes an ability to recognize the ubiquity of emotional experience, a grasp of the complexities of mood, and an understanding of both the mental and physical aspects of emotion, will set us on the path toward self-knowledge. Similarly, Kant argues that we are morally obligated to respect ourselves and others. Respect requires that we promote our own rational being and the happiness of others. In this properly robust sense, it requires understanding and meeting emotional needs. Many theorists of emotion tell us that emotions relate to values; the next step, after we have recognized and understood the emotions that we are having, is to evaluate those values. We turn to that task in the next chapter.

3

Emotions, Decision Making, and Morality

Evaluating Emotions

Emotion is a popular topic, and there are a number of theories that connect emotions to morality and our decision making. Unfortunately, few of them start from an understanding of moral theory and so few can advise us in emotional evaluation in any thorough-going, robust sense. Much of the recent interest in emotion and morality comes from scientific disciplines, and while psychology and neuroscience are, of course, indispensable endeavors, being scientific and observational practices, they are not well suited to presenting arguments about the way that people *should* behave or about the way we should evaluate our emotions. We shall also see that, far from being the external, unnatural, and awkward constraint those unfamiliar with the meaning of morality fear it to be, moral decision making is internal to the experience of emotion, and denying or insufficiently developing our moral nature frustrates our emotional lives.

Emotion and Morality from the Scientific Perspective

Scientists have a lot to say about emotions and morality. From observing our bodies (inside and out) and our behaviors, we can glean much useful information about human nature. In fact, it is impossible to understand emotions (our own or those of others) if we do not have a fair amount of scientific information about them. Nevertheless, because scientific theories are based on observation and logic, when it comes time to evaluate our emotions, they tend to posit spurious values that are based on these observations and scientific theories (like evolution or behaviorism). Nevertheless, just because scientists are not moral theorists does not mean that science and moral theory cannot dovetail in a way that respects the fundamental difference between observing the way things are and reasoning about the way they *should be*.

Patricia Churchland, in her book *Braintrust*, lays out the "neural platform for moral behavior" in terms of our natural caring emotions.[1] She reviews the neurochemical correlate (oxytocin) for feelings such as maternal affection, romantic love, and trust. She also explores the neurophysiological dimension of learning and problem solving. Correctly, after weighing possible evidence for and against a neuro-evolutionary case for universal moral laws, she concludes that no such case can be made. Indeed, given the logic of evolution one must begin with (at least near) universality, and in the case of moral laws, we would need to see an almost completely moral world in order to conclude that there is an evolutionary universal moral law.

While Churchland's project is to use neurology to explain morality, she does not consistently use the term morality the way in which moral theorists use it. In her analysis, "moral" can mean at least three different things: (1) caring and cooperative feelings and behaviors, (2) group practices, and (3) the possibility of universal guidelines for decision making. "Morality," in its most common use, refers to only the third meaning: guidelines for behavior, having to do with notions such as right and wrong and good and bad that transcend individual feelings and group practices. While we mostly assume that caring and cooperative feelings and behaviors are a good thing, from the perspective of moral evaluation, we cannot necessarily make this assumption. For example, taking care of one's friends and relatives cannot be used to justify any possible decision. To serve one's own family by actively hurting others, for example, is immoral. This fact is perhaps even easier to see with group affiliations. Indeed, Churchland's true conclusion is that there is no such thing as morality in the third sense. Unfortunately, there is little reason to credit her with more than begging the question since she begins with the assumption that the notion of morality is exhausted by the first two senses, and her discussion of moral theory reveals a bias that prevents her from giving it honest consideration.

Additionally, there are cases in which our "moral" emotions will need to be evaluated, and we will need to call on higher values. Churchland seems to believe that the theory of evolution yields moral values. When explaining that people sometimes must solve problems with reasoning that goes beyond natural moral feelings, she gives the following example:

> In small groups of hunter-gathering humans, such as the Inuit before the twentieth century, it was not uncommon to capture women from

> the camps of other tribes, undoubtedly a practice that served, albeit without conscious intent, to diversify the gene pool. Those in modern western societies who favor a rules approach to morality would likely condemn the practice as showing a violation of rules. I would not find this an easy judgment, however, since the alternative—in-breeding—has hazards I would not want to have visited upon the Inuit, and would not have wanted for myself, were I an Inuit living in the Arctic in the pre-European period.[2]

Although Churchland objects to "a rules approach" to morality, she intuitively employs its basic premise, as one can see in this example, of putting oneself in the position of the other—"we cannot argue that something is right for me and wrong for you simply because 'I am me and you are you,'" she writes. "There needs to be a morally relevant difference between us."[3] Nevertheless, her ability to morally think through this example is clouded by her assumption that in addition to the imperative for us to put ourselves in the position of the other (the woman being raped), there is also an "evolutionary imperative" to propagate our genes. According to such an imperative, one doubts whether Churchland would ever be able to condemn rape (especially since she likens human morality to animal morality and there is no such animal offense). Her tendency to validate such an evolutionary imperative is so strong that she overlooks the fact that there is no real conflict between diversifying mate selection and respecting women. It is probably the case that prohibiting kidnapping and rape would yield more, rather than less, intertribal mating in this case. Nevertheless, even if one did have to choose between having no children and rape, one should clearly choose the pragmatic loss, having no children, over the moral wrong, rape. In no way does evolutionary theory about the origins of physiology yield moral dictates about the way we should behave.

Joshua Greene, in his book *Moral Tribes*, does a better job illustrating the relationship between science and moral theory. Greene argues that humans have quite a lot of natural moral feelings and behaviors, and his treatment of current research in moral psychology—too detailed to cover here—is quite broad and enjoyable. Nevertheless, Greene argues that our feelings of sympathy and our cooperative inclinations are not naturally universally considerate. They spring from our relationships to family members and group affiliates. It is not hard to see that we feel sympathy more readily for those we perceive to be similar to ourselves. Churchland, without considering behavioral research, stops there and, drawing from a (confused) application of evolutionary theory, is instead tempted to conclude that real altruism, or generosity directed to people who are not likely to reciprocate, is *impossible*.[4] Greene instead correctly points out that not all physiology and behavior are the result of direct evolutionary pressure. Evolutionary psychologists use the word "spandrel" to refer to traits that are *byproducts* of other genes that were themselves directly the result of evolutionary pressures. The whiteness of teeth is a common example: by itself, this color was not a target of evolutionary selection but is a byproduct of the fact that hard minerals, like calcium, are white. Humans have the ability to think complex thoughts (including disinterested moral thoughts about the way the world should be regardless of whether or not I benefit from it) and have complex emotions. Human intelligence does promote human survival in many ways, as through technology and social cooperation, and so human intelligence was likely the result of pressures of natural selection—although, given the fact that nonrational animals survive just as well, we must keep in mind that random genetic variation, sex selection, and fitness to particular environmental pressures have also played a role in human evolution. Given that our brains got bigger and bigger, other things followed, as effects: childhood lasted longer, we made clothes and houses and so could survive with less fur, human society became immeasurably complex, and we gained the ability to reason morally about right and wrong. Given these developments, pace Churchland, humans are in fact capable of sacrificing themselves for others, even if those others are complete strangers.

Greene refers to Wittgenstein's metaphor of kicking away the ladder of evolution.[5] In other words, it is sometimes interesting to wonder, biologically speaking, where morality came from, and accounts like Churchland's or Greene's help to tell that story. Nevertheless, where you come from and *where you are going* are different stories and the former does not constrain the latter. This is a good thing because, as Greene points out, natural moral feelings leave a lot to be desired: they can be partial, bigoted, closed-minded, or just plain wrong. Greene, for his own part, acknowledges that moral theorizing is necessary to adjudicate between rules that might be accepted by a particular group or "tribe" and rules that are fair to everyone and with which everyone can, at least in principle, agree.

Permit me to add one final comment about neurology—although I shall soon discuss Damasio's more sophisticated theory, which is less susceptible to this criticism. Churchland might object that moral theorizing is mere fluff compared to the hard science of brain observation. She seems to assume that brain chemistry is more than a mere correlate to thought and behavior: perhaps she would say that it is the *ground* or fundamental reality of thought and behavior. While a philosopher with a bigger brain than mine will have to solve the problem of determinism and causality between the brain and the mind, we can at least mention here the fact that brain science cannot exist without conscious observations of mental and physical behavior.[6] Without correlations between psychological or behavioral reports and brain scans, neuroscience cannot get off the ground. Given this necessary cooperation, neuroscience cannot conclude that something we experience consciously, like respect for the moral law, is impossible.

"EMOTIONS ARE RATIONAL"

Greene holds that emotions are quick and imprecise reflexes that serve a social purpose, but since they are often too imprecise, we need to use reason—as slower process—to correct their mistakes. In being willing to criticize emotion in this way, he is going against the grain of current psychological and philosophical trends in emotion scholarship, which posit that emotions are rational in important respects. Most people who study emotion today—whether in the affective or cognitive camp—believe that emotions are good precisely because they are values and prompt value-based actions. Nevertheless, these theories are insufficient because they fail to grasp that emotions are not fixed, pretheoretical evaluations but are themselves informed by our thought processes. Emotions are evaluative, and we must in turn evaluate the values which with they present us.

It is becoming popular to defend the "rationality" of emotions, not by assimilating them into the rational mind but by instead arguing that emotions *take precedence* over the rational mind. Two related lines of argumentation have become common in linking emotionality to morality in this way. They are both based on the idea that emotions help unemotional reason make decisions. The first is that emotions give us information about our values. The second holds that emotions aid reason by telling it what is important. Both of these approaches seem to assume that reason and emotion are separate processes and that reason is itself value-neutral.

Murdoch, Nussbaum, Blum, and Walker all advance the idea that emotions help us with the problem of salience, or, since, according to these thinkers, emotions reveal values, they help us to know that some things are more important than other things. Thus, emotions aid us in decision making, especially moral decision making.[7] Yet, the fact that emotions express personal values does not mean that they express moral values. The "information claim," as Stocker and Hegeman put it, represents a confusion about what reason, as well as moral theory, *is*. Simply to know that someone values such and such does not tell us anything about what that person *should* value. Consider, for example, the disgust that someone might feel seeing two men kiss.

This emotion reveals that he or she values heterosexual relationships and does not value homosexual relationships. That information seems pretty obvious as well as useless in the face of any moral questions we will likely have following this emotion.

Stocker and Hegeman examine the evaluative dimension of emotion in their book *Valuing Emotions*, proposing to go beyond "the information claim" that emotions are demonstrations of or give information about personal values.[8] They argue that emotions are not merely instrumentally useful because "emotions are also essential constituents of life and value."[9] On one hand, by this addition they mean no more than to assert that human life is essentially and necessarily emotional. On the other hand, Stocker and Hegeman flirt with emotivism, even though they eventually reject the emotivist claim that emotions are "internal to value."[10] They argue that emotions teach us three things: the value of having emotions, emotional adeptness, and the content of other people's emotions.[11] This idea about learning from our emotions is very much like the notion of emotional universalism I develop here, and Stocker and Hegeman similarly suggest that there are correct and incorrect modes of emotional engagement. Drawing from a connection between the Aristotelian notions of phronesis and habit, they argue that correct emotional engagement reveals the correct values.[12] This seems to be (tautologically) true, but they do not venture to tell us anything about the nature of correct emotional engagement or correct valuing—doing so should be acknowledged as a difficult task that would presumably take us out of the realm of a theory of emotion and into moral theory. Nevertheless, we shall find that when we take emotional experience seriously, moral theory is internal to moral experience without resting on the conviction that emotions simply are values.[13]

Damasio's brain research presents, perhaps, the most sophisticated version of this type of account of the role that emotion plays in moral decision making.[14] He studies people with frontal lobe brain damage who, although they behave normally in other ways, behave in some irresponsible ways that are inconsistent with their pretrauma personalities. They have trouble managing their finances, getting to work on time, and following through with required job tasks. They violate social conventions, sometimes breaking laws, and show a lack of empathy.[15] Previously these behaviors had been explained as an impairment of reasoning or memory abilities, but Damasio argues that these explanations are not satisfactory: he speculates that the problems are caused by a breakdown of emotional, not cognitive, functioning. He argues that the patients fail to bring to mind the appropriate emotional memory that would help them behave properly or make an effective decision. Damasio uses this idea to argue that feelings are "rational," that is, they are beneficial to good reasoning. His argument is similar to de Sousa's defense of the "rationality" of emotion, which holds that emotions help us both to answer questions of salience and to break stalemates in rational decision making.[16] Damasio's theory is a part of his "somatic marker hypothesis" of decision making.[17] It holds that emotions are "integral to processes of reasoning and decision-making" because the "mechanisms of reasoning" are normally affected by "signals hailing from the neural machinery that underlies emotion."[18]

Bechara's work in conjunction with Damasio, especially in the article "Deciding Advantageously Before Knowing the Advantageous Strategy" helps to explain the sense in which they take emotions to be rational. This article describes an experiment in which people with prefrontal brain damage (and decision-making defects) and people without brain damage were asked to play a gambling game in which certain choices were riskier than others. The people without brain damage, after playing for a while, had a hunch that certain decks of cards were riskier. Many of them, after playing for even longer, could articulate the reason that those decks were riskier. The entire non–brain-damaged group avoided the risky decks. The group with prefrontal brain damage did not report experiencing a hunch, even though later, almost half of them had conceptual knowledge of the reason that certain decks were more or less advantageous. Surprisingly, no one from the brain-damaged group avoided the disadvantageous decks, even those who explicitly knew that they were disadvantageous. The experiment, they say, like the accounts of the patients' lives discussed above, demonstrates a gap between conceptual knowledge and behavior. Bechara et al. conclude that human decision making involves two possible paths: affective and cognitive, the latter borrowing from the former. They speculate that the affective path makes use of nondeclarative knowledge that draws on memory of rewards and punishments. Damasio refers to this as a "gut feeling."[19]

Why did the brain-damaged subjects not choose the safer decks even after they understood that they were safer? The fact that those people without a hunch, or, in other words, with a feeling impairment, did not act on explicit conceptual knowledge suggests that cognitive decision-making processes rely on affect, just as they rely on other types of information, not that there are two distinct processes. Furthermore, the fact that there can be a "hunch" that precedes cognition seems to demonstrate the ways that feelings are themselves conceptual. Indeed, this finding that feelings can reveal cognitive insights more quickly than cognition can is interesting and important. Still, there seems to be no evidence to posit a duality of reasoning systems, but perhaps only an impaired reasoning system in the brain-damaged patients. The authors' recognition that somatic markers might be generated cerebrally acknowledges this possibility of a more unified explanation of brain processing.[20] It would be a great shock to the moral theorist of any stripe if Damsio and Bechara could prove their assumption that rationality, acting alone, cannot discover value.

Damasio and others who draw from his research use the term "moral sentiments."[21] Yet, Damasio does not attempt to distinguish between cases in which we should follow natural sympathy or group rules from those wherein it would be immoral to do so. As a biologist, Damasio does not pretend to be a moral theorist, but he does seem vulnerable to the temptation to explain and justify morality and human moral organizations through biology and evolution.[22] Damasio researches the role that brain anatomy plays in the experience of certain moral sentiments, but his research has been taken as implying conclusions for moral theory. In one experiment, brain-damaged people were more likely to make hypothetical moral decisions according to utilitarian values than those without brain damage, who felt more beholden to the value of an individual's life. Theorists, like Philip Kitcher, try to draw conclusions from this about the validity of utilitarianism and Kantianism.[23] Thankfully, they conclude that it is better not to have brain damage! Nevertheless, someone could easily draw the opposite conclusion, as it is just as easy to argue that sometimes normal is bad.[24] Perhaps their brain damage has led them to find a higher moral truth; without moral reasoning, not mere biological observations, we cannot say.

Most current adherents of "the information claim" acknowledge that there can be bad, as well as good, emotions, but they do not attempt to tell us how to judge between the two. If emotion helps with moral decision making, then moral decision making must in turn help us to have good emotions. Therefore, we must let a phenomenology of emotion lead us to ethics, provided that ethics is already reasonably knowledgeable about emotions and open to learning more. In taking this next step, we shall see that emotions and moral deliberation are closely intertwined and we cannot therefore succeed in one dimension without also engaging in the other.

A CRISIS OF PRACTICAL REASON

Many theorists from both the cognitive and affective camps are united in the assumption that the emotions are valuable and should be heeded. Nevertheless, neither an anti- or pro-emotion stance necessarily promises to yield insight. We must reject both the idea that emotions are problematic, unthinking impulses and the idea that emotions are privileged sites of information. We shall see that, moral deliberation is internal to the experience of emotion, so if we aim for moral self-improvement and to be a good person, we must roll our sleeves up and get emotional.

Emotions are often thought to be irrational (not just foreign to rationality but actually opposed to rational thought) because, as with anger, they sometimes seem to cause us to act before we have a well thought-out plan for the action. Imagine, for example, that a father is tending to his four-year-old daughter at a public play yard. The play yard is busy, and among the children are some teenage boys running around carelessly, playing tag. Among them is a particularly tall boy, wearing boots that are easily as large and at the same level as many of the younger children. The father becomes worried that the young children might get kicked, and when one of the older boys inadvertently hits one of the children the father leaps up, as if to attack, with the goal of making the older boys leave the play yard. A verbal fight ensues, but the boys eventually leave. The father then feels vindicated, but also regretful, and wonders whether his actions were justified and optimal.

Perhaps at this point he, or others, might blame the "emotion" of "anger" for causing him to act so attackingly instead of coolly addressing the boys and making them aware of the problem. Some affect theorists will point out that the action was a product of the "fight" response and an automatic result of the adrenaline coursing through his veins. On this model, the individual must find a way to control the anger after the fact since the response itself is automatic. A functionalist account will adopt this same explanation but with a positive spin: if it were not for the feeling of compulsion, the father might have done nothing, and someone might have gotten hurt. Perhaps the "fight" response was warranted given the prevalence of violent encounters in our society and the appearance of the boys. Greenspan argues that emotions are "rational" because they are quick and useful impulses, but in this situation it is exactly the pragmatic and moral evaluation that is left undetermined.[25] Did the father do the right thing? Even he wonders.

Again we are faced with the choice between thinking of the "emotion" as either bad or good. This dualism misses a couple of facts about the situation. First off, it is not likely that the father saw the boys as a threat in the same way that a caveman sees a mountain lion as a threat: if that were the case, he likely would not have let his daughter play there but would have swung her over his shoulder and ran.

Second, the father was worried before he was angry. Perhaps he was worried that his daughter would get hurt; perhaps the worry was a result of internal strife and fear of confrontation. If we look at it this way, we see that the altercation with the boys, or the anger, was a brief episode in the life of the emotion, which was, for a much longer time, preoccupied with the question: What should I do? What should I have done? The father was asking himself pragmatic and moral questions. What starts out as an example of an emotion reveals itself to be an example of moral deliberation.

Might the moral decision making have gotten on better if the emotions did not exist? Might a Spock-like moral reasoner have acted differently and more effectively? (Should our slogan for emotional deliberation be WWSD?—What would Spock do?) Most people now answer this question negatively, but that is because they believe that Spock would fail to see the moral and pragmatic urgency in the first place. I think that that is wrong. I think that Spock might have done a better job, depending on how much Spock knows about human nature and moral reasoning. If we ask the man now, after the emotions have subsided, "What should you have done in that situation?" he will most likely say, "I don't know." This fact is important. He might come up with some hypotheses, but nothing about which he feels sure enough to say that he will try to do next time. Spock would only be a good guide if Spock is morally committed and well versed in human psychology. Since the fictional Spock frequently voices puzzlement over human behavior, he likely would not have been very effective in dealing with this situation, but maybe after a brief session of coaching in conflict resolution, Spock would have done a great job. Unfortunately, brief coaching sessions in conflict resolution are not easy to come by.

Any person in this situation needs both knowledge and practice. The father needs to know what he should have done and the only way to be *certain* of that knowledge is to have had practice dealing with similar situations. Imagine a dad who has himself a teenage son *and* a young daughter. If he is himself good at managing their relationship—perhaps if he himself had a good role model or if he read a good book on this topic—he will think that the father in the above example simply lacks knowledge. Our world would be a wonderful place if the majority of my readers found this example unrealistic, but my sense is that most of us lack the basic knowledge and experience of conflict resolution to make that the case.

Ask yourself: what should the father have done? You might say that he should have calmly and respectfully approached the boys, making them aware of the problem, letting them know that he was feeling worried and apprehensive. He should have acknowledged not only the boys' desire to play, but also the right of the children to play, and been open to finding some kind of compromise. Now ask yourself: how many people that you know have the *courage* to do this? Does the mere idea of telling a stranger how you *feel* make you roll your eyes? Most of us do not want to engage with strangers. We are afraid of making ourselves vulnerable or overstepping our rights; we assume that the other person will harm us in some way or simply that we will fail to make ourselves understood. It is much easier to demean and retreat than to muster the moral courage to respectively interact with others. My guess is that most people would just leave or tell the child that she couldn't play there anymore. (As a parent, I am surprised at the frequency with which I see other parents demand that their children capitulate to the will of their peers so that they can avoid a potential conflict.)

Perhaps not all emotions can be called "crises of practical reason." It would be odd to describe *positive* emotions as "crises." Nevertheless, much of human psychology involves decision making, and any account of emotion that easily identifies emotion with morality does justice neither to morality nor to emotion.

LAYERS OF EMOTION

Emotions are hardly ever simple. They are usually complicated by other emotions and our emotional evaluations of them. In fact, our evaluations of them are often much more complicated than the emotions themselves. Humans can take a simple emotion, like joy or sadness, emotions that many other animals exhibit, and turn it into something totally unrecognizable and incomprehensible, even to other humans. Our implicit theories of emotion, like Augustine's theory of sadness as a sin, influence our experience of emotions. This fact is the reason that it is necessary for even the most unphilosophical person to have a handle on a good theory of emotion (and perhaps the reason that same task is philosophically so difficult). The basic emotions are themselves trivially easy to understand; often, discussion of "the laws of emotion" or "core relational themes" tells us nothing more than the meaning of the words we use for emotions; for example, "sadness" tracks "having experienced an irrevocable loss."[26] Nevertheless, if we look at an authentic, lived experience of sadness, we will see anger, violence, drug use, laughter, fun, even philosophy... These layers of the emotion are a part of the context that must be taken into consideration when we try to understand our emotions, evaluate them, and decide what we should do about them. The fact that lived emotions are so complicated, often repressed, and related to other emotions suggests that emotions are self-evaluative.

Emotions call for attention and expression, and, in that way, emotional experiences seem to involve crises—the Greek word *krisis* means "decision"—of practical reason: they demand: "what should I do?" "What can I do?" or perhaps "What is the meaning of this?" In the case of negative emotions, whose affects usually begin internally, the question is more, "What should I do now?" In the case of positive emotions, whose affects are sometimes expressed externally more easily, the question "What will I do?" is a call not only for immediate action but to also reevaluate one's commitments and identity and to plan accordingly.[27] Hence, as current theorists of emotion argue, emotions let us know that something is important, but further thought is necessary to determine what we should do about it. Emotional situations are important not because their meaning is given, but precisely because their meaning is contested.[28]

We have seen that emotional experiences can themselves be concerned with moral questions. As de Sousa remarks, judging whether or not an emotion is "rational" is a "complicated process [that] is at the center of our moral life."[29] We must keep in mind that even the evaluation of emotions is emotional. (Kant's theory of emotion as the physical feelings that accompany certain thoughts does a good job of keeping thought and emotion linked.) The father in the above example might have felt some degree of self-contempt during this episode; similarly, thinking about moral questions might make him fearful. Especially with negative emotions like anger, we often feel that we should not feel them while we are feeling them (and hence feel some type of guilt or frustration). Of course, this self-evaluation may or may not be morally correct.

Thereby, it would seem that "emotions are judgments about judgments." In other words, emotions not only contain evaluations, they are self-evaluative. Prinz tentatively attributes this "meta-cognitive" insight to Nussbaum, even though he eventually expresses it himself: "To assent to a value laden appearance, one must form another judgment, to the effect that this judgment is justified."[30] Kant also expresses this insight with regard to desire. In my mind, the fact that emotions are inchoately self-evaluative entails the necessary relationship between the experience of emotion, theories of emotion (as we saw in the first chapter), and moral theory. The last two ask and answer questions about which emotions we should feel, while emotions themselves prompt these questions while provisionally assuming answers.

In the case of negative emotions, it is more often the case that we form the meta-judgment that the emotion is not justified. A good example of this comes from St. Augustine's *Confessions.* I gave this example in chapter 1, but I will repeat it here. After his conversion, Augustine stoically judges that the feeling of grief is a sin since it reveals an attachment to the material world. With the death of his mother, he wills himself not to feel sad; he succeeds in only crying a little bit. He then uses sleeping and bathing to change the emotion. Instead of sadness, he is next overcome with anxiety over the fate of his mother's soul, and pleads with God to save her, even though he admits having no reason to worry. The unacceptable sadness was changed into a "devout" sentiment, while the physical feeling was retained. In this case we can see that the evaluation of the emotion changes the emotion itself, even in a way that exceeded Augustine's conscious control and awareness. Moral theory was called on to evaluate the meta-cognitions that accompany and inform emotional experience and, in this case, to therapeutically alter them. Unfortunately, Augustine's Christian evaluation of and experience with death is a common, yet flawed, model for sadness.

Perhaps we are more likely to repress negative emotions (and sometimes even positive emotions, especially to their full extent) because the needs they express remind us of our finitude. In the language of psychoanalysis, they belie the wish (of the ego-ideal) to be whole and self-sufficient. Freud writes that repression comes from the ego-ideal,[31] and that feelings of guilt and inferiority accompany negative emotions.[32] Nevertheless, emotional self-understanding involves accepting emotions, which means accepting our own weakness, dependency, and finitude. In the upcoming chapters, we will further discuss the foundational value of transparency and honesty in the pursuit of emotional and moral self-improvement. Here we will begin to discuss the ways that moral decision making bears on these crises of practical reason.

KANTIAN EMOTIONAL EVALUATION

Kantian moral theory can help us evaluate our emotions. In this section I give only a brief summary of Kant's moral theory; in the following chapters (and in the Introduction) I explain his main theses more fully. Here we must ensure that the reader understands the terms, "morality" and "good," we have been using. Then we can precede with our discussion of the relationship between emotional evaluation and moral evaluation.

Moral reasoning is itself the ground of value, for Kant. This fact will be a relief to those who do not feel comfortable with a moral code based on God or religion, but it might seem strangely circular nonetheless. It expresses the intuition that we do not normally take moral considerations to have the same weight as pragmatic considerations: for example, we do not weigh how much we want some child's ice-cream cone against the negative value of punching that child to get it. Weighing moral value against pragmatic value is itself immoral.[33] Similarly, you cannot weigh moral concerns against money-making concerns; moral concerns trump all other concerns.

The notion of inherent value is special to Kant's philosophy. It means that, contrary to something that has value because it is useful for doing or getting something else or because it is pleasurable, some things have absolute value regardless of whether or not they are profitable, useful, or pleasurable. Morality is the means of figuring out what has absolute value. Morality itself thereby has absolute value. Similarly, the fact that human beings are capable of moral reason makes them inherently valuable and deserving of respect.

Kant's moral code is usually explained in terms of "the categorical imperative." The categorical imperative is the moral law that itself contains the moral code. Although there are different ways of expressing this rule, all of its different expressions boil down to the same thing, which is the inherent value of moral reason. The first formulation tells us to act only in ways that are not predicated on our getting away with something. Some behaviors only work because we are tricking others in some way: when we lie, we expect that other people will think we are telling the truth; when we steal we expect that other people will respect our property. These ways of coercing and manipulating the expectations of others contravene the fundamental nature of reason, namely that it is based on universal, transparent rules or concepts that everyone can understand and agree to.

Although the "universal law" formulation of the categorical imperative is the first one Kant explains, the second formulation, the "respect for rational persons," is somewhat easier to understand and perhaps more basic than the first. The second formulation of the moral law is that we should always respect rational personhood. We should respect ourselves and respect other people. We can all think of examples of what this respect entails as well as ways that this respect is relative to different contexts. In this regard, Kant tells us that, in formulating our goals in life, we have the responsibility to promote our own self-perfection and the happiness of others. It is interesting, perhaps mysterious, and important that we do not have the moral duty to promote our own happiness; nor do we have the moral duty to perfect others. We all too often exchange goodness for our happiness, and we have to constantly be on the look out for this mistake. Doing the right thing is seldom easy; unfortunately it can also be unpleasurable. We do seek our own pleasure—that is inevitable—but it is evil if we put the pursuit of happiness above morality.[34]

In chapter 5 I discuss these tenets in greater depth, but here it is important to defend Kant's perfectionism. Perfectionism is the position that it is our moral duty to strive to be, not just acceptable, but perfect. Most people resist moral reasoning because of the self-criticism it involves. If we admit that there is more that we could be doing, it seems that we have to live with the sinking feeling that we are never good enough. First and foremost we must accept that we are not perfect and that we could improve. To give up on self-criticism and self-improvement is to turn away from the pursuit of goodness, as well as to evade the true purpose of life. Moral commitment involves courage and a strength of character: we must accept that the world is a flawed place and still be committed to making it better.[35]

With this brief orientation in moral theory, we can return to our consideration of emotion. We can see that the values involved in emotions are often related to moral values and moral deliberation. These general moral guidelines do not specify exactly what one should do in any specific circumstance. To do so would be to offend our rationality. Instead, moral deliberation is a constant fact of life and closely related to our navigation through emotional experience.

If we employ Kant's moral code, we can evaluate the behavior of the father from the example above as well as gain some clues about how he should have acted differently. First off, it is not morally acceptable to give commands, such as "Get out of here!" To control and intimidate others is not respectful and offends their ability to reason for themselves.[36] Contrariwise, invoking common rules (universal reason) sometimes helps, as long as those rules are mutually acceptable. Similarly, referring to fairness can also help. The father could have said, "This play area is recommended for 3–5 years olds, so it's not fair for you guys to make it hard for the little kids to play." Of course the older kids might have said, "No one follows those guidelines." Or, similarly, they might have called presumed rules to their aide too: "The sign says that kids need to be supervised" or "We're just playing tag." This kind of rational negotiation that refers to common understandings and values respects the rationality and personhood of both parties. Yelling and trumping another person's judgment cannot be made into a universal law; it is not even acceptable to the most immediate party. The father certainly does not want the boys to treat him that way in return. Furthermore, we have to think about universality—the type of culture and world our actions theoretically create. (The duty of the parent to educate makes these moral exercises more salient.) In the terms I used earlier: Is the father trying to *get away with something* by his expression of aggression? Can we will a world in which parents do *whatever it takes* to protect their children? Or teach lessons of security by means of threats and intimidation? Kant's method of testing the universality of our maxims is quite informative in this case since such a world is not even hypothetically possible. We cannot make a more peaceful world by threatening violence. A world in which people cared for nothing other than their own children would similarly be a violent, unsafe world. It does seem that the father is making an exception for himself; he does not want the boys to respond in kind. Similarly, it seems that what he is really denying are his own feelings of insecurity.

Nevertheless, all of our grand thoughts and reasoning seem to go out the window at the first sign of our actually needing to use them. In the event that we endeavor to therapeutically alter our emotions, it is likely that we will need to employ both rational insight and behavioral training. The above "talking through" of what to do might help, but the rational insight will necessarily need to include a thorough understanding of the causes of the emotions. It is not effective to demand change before and unless the reasons behind the behaviors have been well understood. As we will discuss further in what follows, emotional insights are often moral insights; to ignore them is to ignore morality. It is likely in the above case that the father's worry was a species of fear and that he was embarrassed of that feeling and unable to express it directly. Asking for help from another adult at the playground, for example, would have also likely created a favorable outcome for him. His own emotion (of fear) is just as important in addressing the moral demands of this situation as is addressing the feelings of the boys.

In this example, it matters little whether or not the "emotion" is "rational." Maybe the child was not safe, maybe she was. What matters is whether or not the intentions and behavior are morally acceptable. If the father's anger was calculated to overcome his fear of conflict, then we might conclude that the father's emotion was based on a false belief: the father is, *in theory*, capable of dealing with this situation. The emotional situation also led to a less than desirable outcome. Nevertheless, both the belief and the affect are based on a very real lack of experience and knowledge. He cannot change that belief without either gaining that knowledge or irrational courage. Perhaps his beliefs are even based on the fact of past failures and fights. Very few of us are any good at confronting and changing the behavior of strangers. We must not fall into the trap of thinking that "reason" is omniscient and omni-benevolent. It is only prejudice that makes the question about the degree to which emotions are "up to us" look like it makes more sense than the degree to which thoughts are similarly "up to us." Furthermore, calling emotions judgments, while partly illuminative, gives the false impression that emotions are fully self-transparent, with relatively fixed boundaries. The fact that emotional thoughts can be unconscious casts doubt on this assumption. Behind it all, as we saw, is moral deliberation, and more fully engaging in that moral deliberation can help resolve the questions that are internal to the emotions.

We may decide to try to ensure that future conflict situations do not play out in the same way. It is difficult to say whether cognitive or behavior therapies would be called for in any particular case. It seems to be more preserving of the respect of the individual to err on the side of cognitive therapies, but that approach runs the risk of excusing bad behavior. In either case, having a good understanding of emotion and moral psychology will help us change.

UNIVERSAL AND PARTICULAR

Moral theory is internal to the experience of emotion to the extent that it is internal to experience in general. Yet, most models of the subjective encounter with moral theory portray it as a form of abstract deliberation. Then one asks what role emotion should play in these thought processes. It is thought that emotions are necessarily partial and hence resist the impartiality of moral reasoning, but this concern continues to assume that emotion and reason are two different and unrelated mental faculties. If we can show that emotion and reason are internally related, then the role that emotions play in moral reasoning will come into clearer view, as will the compatibility between universality and particularity.

People seem to be easily confused about the way that we are to morally evaluate our special, personal relationships. It is thought that the mere fact of special, personal relationships contravenes universality. Not everyone should love my husband and children, nor do I owe special regard for other people's husbands and children. At the same time, not everything that I might do in the name of love for my husband and children is morally permissible. This seems obvious, but when we add in the talk of "universality," people get confused.

Perhaps the confusion involves the proper way to express our principles. We do not ask whether or not it would be okay for everyone to go to the zoo on Saturday with my kids: the zoo is not big enough to hold everyone. We ask whether or not anyone *in my situation* can do what I do.

Similarly, within Kant scholarship there has been a recent spate of attention to "context" and "particularity" and the relationship between the subjective and objective in ethics, along with practical wisdom, which is said to be necessary to bridge this supposed gulf. Nevertheless, a thorough understanding of Kant's theory of virtue shows us that there is no such gulf. Herman argues that Kant's theory is based on the assumption that we already have basic moral knowledge that we bring to bear on any situation.[37] For example, there is a lot of psychological know-how that the father on the playground would need to pick up in order to put his newfound moral insights into practice. For example, he would have to use the right tone of voice when speaking to the boys. Not many people have tried to apply moral reasoning directly to the mundane questions of everyday psychology. When we try to do this, as we did above, there does not necessarily seem to be any practical question into which we cannot go. Of course, sometimes our attempts will simply suffer from lack of (scientific) knowledge. For example, maybe the older boy in the example above had some kind of disability, like muteness, about which the father was totally ignorant. Without relevant knowledge about the facts of the situation, one will not be able to judge what to do in it. Nevertheless, that fact does not lend itself to the conclusion that there is a gulf between universal evaluation and particular contexts; it entails only that we do not know enough to make a good judgment.

EMOTIONAL INSIGHTS

Moral psychology can also help us to understand the emotional dimension of life that we need to respect. Just as we do not create but work with the laws of logic and the facts of experience, the narrative we construct in coming to understand our emotions makes use of relatively fixed, or natural, psychological laws. Nevertheless, feeling, thinking, and self-alteration are necessarily interconnected parts of a whole life: trying to understand one in isolation from the whole necessarily yields a distortion.

Those thinkers who argue that emotions express and are related to values are not wrong, and sometimes it is the case that emotional values must be affirmed and strengthened. Indeed, feelings appear to come from a different place than reason does, and this apparent difference can confuse us. We can make an analogy with thunder and lightening: we are not born with the knowledge that thunder is the sound that lightening makes just as we are not born knowing that a tightened stomach—or whatever feeling a particular person experiences—is the feeling that being belittled makes. Coming to understand and expect their identity is necessary for emotional literacy. Furthermore, it may appear that Damasio is right and that emotional values shape life decisions in a way that, what he calls, "pure reason" cannot.

When my dad died it seemed weird to me that the sadness kept coming back. Feeling very strong emotions in the weeks following his death made sense, but why should I feel sad every weekend for months to come? I was not very close with my dad, and because of his addictions and lifestyle his death was not a surprise. It seemed weird to me that the simple thoughts, "My dad is dead" and "Gloria never met her grandfather," which are sad, of course, lasted so much longer in affective form than in mental form. The possibility of thinking these thoughts lingered, and so did their feelings. (There were other thoughts and feelings, like anger and moral self-scrutiny, that also lasted a long time, but their duration made sense to me.) This feeling of sadness seemed to posit a value of its own, informing and shaping reason; even if I did not think these thoughts, their feelings were there, reminding me of them.

Nevertheless, my sadness was morally good and rational in any normal sense of the term. Reason is a natural function of an organism, and humans are familial animals. It is rational to be tied to one's family. We all need special, life-long relationships to which we are committed. Navigating these relationships helps us to become a better person. If we take reason to refer to selfish, means-justifies-my-end thinking, then moral commitments might look irrational, but we should not employ this limited definition of reason.[38] Moral reason, as Kant explains, is the highest form of reason, and the values it involves are emotional. Perhaps an organism with no emotion would think the thought "Gloria never met her grandfather" one time and then move on, but that organism would not thereby count as more rational, nor is that organism likely to evolve as a part of a species of rational, social animals.

The primary motivation of Prinz's affective theory of emotion is to lay the groundwork for a moral theory. Prinz hopes to harness those emotions that seem to reveal moral truths; we might call these natural emotions, although here the notion of naturalness already implies moral approval. Let us take the example of sadness at death. If someone proclaimed herself to be not upset (the word sadness is not even strong enough) about the recent death of her mother, that she neither had sad thoughts nor sad affects, we should be very worried. We would attest that, in some sense, she actually *is* sad. Based on our understanding of human psychology as well as moral commitment, she simply must be sad. There are very few possibilities that could explain the lack of this nearly universal emotional experience. (This example shows us one way that the emotions transcend both affect and cognition.) Not only is it necessary for psychological health that she feel sadness, but it is also a moral expectation, both to honestly recognize the importance of one's relationships and the identity of oneself, and also to, as is often said, "pay one's respects."[39]

De Sousa defends this objective view of emotion: the idea that emotions perceive real (axiological) properties of the world.[40] Adorno similarly develops an objective theory of emotion regarding aesthetic experience.[41] Prinz tells us that we can conclude that certain events, like death, have moral value because they characteristically elicit certain emotions, but this seems to put the cart before the horse. If we somehow found ourselves among people constitutionally incapable of experiencing sadness at death, we would not thereby conclude that death was morally meaningless. Aristotle tells us that the virtuous person feels happiness and sadness at the correct things because he has been raised in a virtuous society, and from a Kantian perspective even if we lack a virtuous society, moral deliberation can tell us what these scenarios should be.

It is absolutely the case that some emotions are morally good, either because they express proper moral commitments or because they are an expression of our commitment to morality. We will explore Kant's theory of natural moral emotions in the next chapter. While, in the style of Freud's libidinal theory, we might grant that emotions bear a special relationship to instinctual drives,[42] in opposition to Freud, I take the moral impulse to be akin to such a drive. For Freud, the instincts are oriented toward love, which is itself a drive for self-overcoming.[43] If we keep in mind that we must understand emotions within the context of natural needs and rational self-development, we will not fall into the Stoic trap of thinking that emotions are so easy to explain away. In other words, a theory of emotion must have a robust appreciation of social and emotional needs, of love, or else the emotions will necessarily appear mistaken or dim-witted.

There are also cases in which our more theoretical moral convictions are repressed and become emotional, lest they be lost entirely. We might take the latter case as an example of emotions taking on the Sartrian "magical" functionality accomplishing that which we do not feel consciously able to accomplish. I can give a personal example of this: once I had a particular topic I needed to discuss with my hair stylist—a sort of apology I felt obliged to give—and yet I had failed to give it during the previous hair cut. Cutting my own hair caused me (unconsciously) to have an excuse to see her again, and it gave me a further motivation to get the apology over with, lest I feel the unconscious need to cut my own hair again. Another example is a time when I fainted during a medical experiment in which I felt that the dignity of my body was being violated. I did not feel I was able to explicitly address the doctors running the experiment and rescind my consent, but fainting succeeded in rescuing me from the situation. These are both examples of repressed moral thinking causing an affect, as if by magic. We have a real, psychological need to follow through on our moral convictions. It is only the portraying of moral experience as any less than "embodied" that leads to the pseudoquestions about the relationship between morality and emotion.

The purposes that my emotions serve are not guaranteed to be morally acceptable. Sartre suggests that all emotions serve disavowed purposes, and, while that should be tested empirically, we are morally beholden to scrutinize our emotions on this score. After all, the father's anger did work to scare the boys away, which is exactly what he wanted. It is also possible that the assumption that anger and confrontation are socially unacceptable caused the expression of the emotion to take the form of an explosion that was calculated to override the internal censors. If that was his true goal, he is morally obliged to acknowledge it.

The examples above all include fairly thorough resolutions. Unfortunately, there are also cases in which our moral/emotional convictions are repressed and transferred into different emotions entirely, such as sadness, anxiety, or aggression. In this way emotional/psychological health is related to virtue. Following through on our moral knowledge and conviction requires courage. Kant characterizes virtue as "fortititude." This can sometimes be the courage to look critically at our selfish emotions, but it can also be the courage to *have* emotions. Our relationships with our closest loved ones provide countless examples of the necessity of this type of courage and fortitude.

We should consider a possible objection to this idea that we must always evaluate our emotions: perhaps it is the case with some emotions that consciously evaluating them would deter the subject from moral ends. What if evaluating emotions undermines their purpose? Greene, in pairing the prisoner's dilemma with evolutionary psychology, suggests that certain feelings, like love, shame, and awe—moral feelings—function in a sociobiological sense like an "emotional straight-jacket."

As with the decision strategy of "mutual assured destruction," what we might call the nuclear threat, these feelings prompt us to override our "best interest" and cooperate. Romantic love, in Greene's example, accomplishes cooperative childrearing or at least the promise of it by evincing a willingness to forgo a better possible mate in the future, hence promising loyalty.[44] A purely rational person would say to her love object: "I want us to agree to stay together forever and work cooperatively, but of course, there is nothing special about me and nothing special about you; either of us could pair up with anyone. Still, we should pair up with each other to benefit from pairing and if we both pair for life, we'll both benefit more, so let's do it, okay?" The love object might go for it, but it seems pretty risky. The emotion of love, on the other hand, (magically) turns the world into the world you need it to be in order to get what you want: it makes you and your love absolutely special and only capable of pairing with each other. Reason creates the emotion in order to transform the world in a way it cannot on its own.

While I find the evolutionary dimension of this argument unconvincing (since there are no corollary universal physiological structures and the feeling must be affected in only very specific situations), it does seem that this account of the purpose of love could support a Sartrian or other disavowal theory of emotion. On such an account, the emotion is affected for the sake of a goal, and the goal can only be reached if the emotion is real, so the underlying intention is repressed. One must further note, however, that there is a certain Catch 22 nature to the emotion—if I am rational I cannot get what I want, but if I am emotional, I cannot acknowledge my motives. The emotion accomplishes what reason alone cannot accomplish.

We might say the same thing for other moral emotions: they help build a bridge between the IS and the OUGHT by means of a hopeful, transformational delusion. The feeling of respect for the moral law, or Kant's notion of a supersensible realm of morality, might be seen similarly as an effective fantasy. This way of speaking is perhaps no different from elaborating an account of the rational moral incentive. In this way, emotions are rational, but they give mere pragmatic rationality an added boost, helping to transform it into moral rationality. This account might help to explain the reason that you cannot have morality without emotion, but you also cannot be immune from morally evaluating emotion. Emotions function in a performative way, inspiring action of which pragmatic reason alone cannot fully make sense.

Can we rationally evaluate emotions on this account? Could the lover step back and ask whether or not her love is *honest* or *good*? I think so. It seems to me that the lover could both become aware of and affirm this strategy, which must be what happens in any case, yet it is repressed. The worry, of course, is that becoming further aware of the emotion would weaken the love and loyalty. If you realize your spouse is not unique and you could pair with someone better, then you would want to, so the objection goes. If you realize that moral feelings exist to make you more likely to behave morally, they may come to look suspect. Nevertheless, divorcing your spouse because you do not love him anymore is no less *honest* than divorcing him because you realize that you never did. The moral emotion is legitimate if you can affirm the underlying strategy; it is *honest* if you make good on it. Since there is no guarantee that any type of self-delusion is in fact moral, there seems to be no reason to grant the emotions that rely on it—if any at all do—a pass on moral evaluation.

WHY MORALITY?

Before moving on to consider Kant's account of our natural moral feelings, we will consider one more likely objection to the idea that emotions must be morally evaluated: isn't it enough to evaluate our emotions simply in terms of our longer term personal commitments? Why bring in some transcendent values that I might not even hold? Gesturing back to the discussion with which this chapter started, we see that when psychologists do discuss emotional evaluation they do not usually make use of moral notions, and perhaps there is some good reason for that. In addition to holding that there are evolutionary imperatives in decision making, it is common to think that humans not only naturally seek pleasure and avoid pain but that they *should* seek out pleasure and avoid pain. While we might ignore Rasmussen's book, *The Quest to Feel Good*, because of his dogmatic devotion to Adlerian psychology, he offers us a good example of the way that psychologists often suggest that we evaluate emotion not by means of moral theory but with whatever theoretical, psychological tools they happen to have at their disposal.

A moral theorist would see Rasmussen as a psychological hedonist, a position that is frequently criticized for being nonfalsifiable. Psychological hedonism is the position that we always act so as to promote pleasure and avoid pain. A psychological hedonist insists on looking at an example of moral behavior—Martin Luther King Jr., going to jail, for example—and pointing out that really the motivation was to achieve pleasure—perhaps the pleasure of feeling that he did the right thing, for example. There is no possible counterexample that could not be twisted to support this interpretation. Being nonfalsifiable, it is not truly a theory and, in this form, does not relate to an argument about what we should do.[45]

Rasmussen also argues that emotions should be evaluated in terms of their ability to most objectively reflect reality and best promote our own personal fulfillment. As a practicing psychologist, he is well versed in the complex nature of emotion and behavior. Nevertheless, if it is true that "humans want to find reasons to feel good about their lives," it would make more sense to simply look for objective good itself rather than personal pleasure or fulfillment.[46] Moral theorists often dismiss the hedonic imperative—the idea that we should only seek pleasure—by means of what is called "The Hedonic Paradox." The paradox involves the likelihood that people pursuing mere self-fulfillment will feel as though life is meaningless. Happiness is not often best achieved directly, but only as a byproduct of other meaningful and important pursuits. This paradox does not directly support the conclusion that morality has inherent value, but it at least causes us to suspect that perhaps happiness does not. Similarly, the therapist who hopes to help his patients find happiness might achieve that goal more directly by asking them not whether or not what they are doing is making them happy but whether or not what they are doing is good.

Those who raise moral concerns are often shunned as being neurotic, and the philosophically minded might be told that they "think too much," but in my experience morally based decision making is often simpler than pragmatic questions about which course of events is likely to yield the most happiness.[47] Combined with the sinking suspicion that such deliberations are ultimately meaningless, time spent trying to figure out what color of new car will be most pleasing or what size television set best suits one's needs, bores and frustrates because the values that undergird such deliberations are ultimately groundless. Comparatively speaking, whether or not one's actions are morally acceptable is an easy question—although one often does not like the answer.

Last, the idea that we need only evaluate our emotions in terms of most efficiently promoting happiness does not do justice to the relationship between emotion and values *that we take to be transcendent.* Similarly, however psychologically astute, these accounts of emotion often leave out a consideration of moral emotions, like respect for the moral law or sympathy. It is to a discussion of this other dimension of the overlap between emotion and morality that we now turn.

4

Moral Feelings

"I always tell my kids to cut a sandwich in half right when you get it, and the first thought you should have is somebody else. You only ever need half a burger."[1] This line from the popular comedian Louis C. K. reminds us that moral feelings are an ubiquitous feature of life, both in our impulses to act and in our response to moral actions.

Kant has plenty of critical things to say about emotions (or what he calls affects and passions):[2] they can be impetuous, obsessive, antisocial, selfish, even evil.[3] Similarly, commentators on Kant have tended to assume that, because he expresses morality in terms of pure reason, he pits reason against emotion.[4] Less often do we hear about Kant's theory of moral emotions (i.e., moral feeling), and it is still relatively scandalous to suggest, as I do, that the topic of moral emotions offers us the best starting point for understanding Kant's theory of emotion. Furthermore, since some readers might be new to the concept of morality and moral evaluation, it is even possible that Kant's theory of moral feelings—not to be confused with moral sense theory—is the best starting place for understanding morality overall, as this chapter will demonstrate.[5]

In his "Doctrine of Virtue," Kant argues that there are two duties of virtue: to perfect oneself and promote the happiness of others. There are a number of virtuous emotions, like respect for the moral law and well-grounded sympathy, that accompany these pursuits. Morality similarly commands that we cultivate any feelings that help us accomplish our duties. Even still, Kant is often critical of emotions because he understands virtue in terms of struggle between morality and selfishness, and it would rightly be a mistake to think that all of virtue could be reduced to pleasant feelings.[6]

In this chapter we will explore everything that Kant means by the term "moral feeling" and the extent to which the experience of moral feelings can help us navigate moral decision making (and vice versa). I will suggest that we should understand moral feeling in an even more expansive sense than Kant explicitly sets out and that doing so will help us to fully grasp the role that morality (already) plays in our lives. We will see that moral feelings play a greater role in our emotional lives than we realize and that we should work (although not without critical self-reflection) to make good on these emotional, moral insights. In the overall task of using Kantian moral theory to evaluate emotion, we must start by recognizing that the verdict of this evaluation will often be in favor of feeling the—or even more of the—emotion itself.[7] Although moral deliberation cannot be conflated with following our feelings, Kant does recommend that we cultivate many specific necessary moral feelings as well as a general demeanor of cheerful virtue, which involves a sensitivity but not sentimentality.

WHAT IS "FEELING"?

In seeking to understand Kant's theory of feeling, we should look to the second book of the *Anthropology from a Pragmatic Viewpoint*. The discussion in that section, "On the Feeling of Pleasure and Pain," is in general agreement with the many comments about feeling that we find scattered throughout Kant's texts. As noted in chapter 2, this is the proper starting point for an understanding of Kant's theory of emotion, not his theory of affects, which will be more fully discussed in chapter 6, because affects are defined first and foremost as feelings.

Gregor argues that while Kant distinguishes between feeling and sensibility (which includes perception), he "moves freely from one to the other."[8] Nevertheless, Kant states the distinction thus:

> We should distinguish between inner sense which is a mere power of perception (of empirical intuition), and the feeling of pleasure and displeasure—that is, our susceptibility to be determined, by certain ideas, either to hold on to them or to drive them away—which could be called *interior sense*. (A 153)

"Sensibility" refers to that part of the cognitive faculty that receives information from the senses. "Feeling" refers to the physiological effects that come from the various operations of the cognitive faculty. In this passage, Kant describes feeling as being like a judgement—an attraction to or repulsion from—certain ideas. Especially given Kant's notion of inner sense as the realm in which all of our sensations (both internal and external) are integrated, it is easy to see that there is a possible slippage between the two notions.[9] Perhaps we see the same kind of slippage in Damasio's work, with his notion of consciousness as "a feeling of what happens."[10]

In the "Doctrine of Virtue," Kant makes the distinction between sensibility and feeling in terms of objective and subjective feelings. In the objective sense, we have a "feeling" of everything we perceive or think. In this case, the feeling is objective because it is *of* either a physical or rational object, and we can presume that this exact same experience, even though it is subjective, could be had by any person. In the second sense, "feeling" is even more subjective: when we have a perception or a thought, that objective feeling can cause a subjective feeling in us. In other words, all kinds of mental events can cause pleasure or displeasure. In addition to seeing a certain color, we can feel an aesthetic enjoyment at the sight of it; in addition to understanding the words of a comedian, we can laugh and feel humored. We don't just *think* a color is pretty or think a comic show is funny; we might not have those thoughts at all, the feeling is instead a physical occurrence. We might let in a small gasp of air and focus our attention on the color, or, of course, we laugh at the funny joke. In other words, (subjective) feelings are bodily events to a greater degree than are thoughts and perceptions. Subjective feelings are the subjective, bodily effects of other, more objective, mental events (DR 211–213).

Feeling, alongside cognition and desire, is one of our mental faculties. In addition to thinking and wanting, we also have the capacity to feel. While it is probably the case that desires can also cause us to have certain feelings, Kant focuses on the feelings that arise from different cognitive operations, namely sensation—he gives the example of amusement or boredom that characterizes immediate experience—imagination, understanding (concepts), or reason (ideas). (Given this explanation of feeling, it stands to reason that Kant thinks that we are perhaps always having multiple feelings, and given that all of those feelings are going on at the same time, we can now see that, in practice, it might be difficult to distinguish as well as see the relationship between different perceptions or thoughts, feelings, other physical sensations, and desires.)

Although the discussion of intellectual feeling in the *Anthropology* focuses on aesthetic pleasure, moral feeling (the pleasure or displeasure that is caused by moral ideas) is the most important and most often discussed feeling in Kant's philosophy. Much of his theory of feeling is geared toward making a place for it. In the very beginning of his *Anthropology*, which is more concerned with what we now call psychology, Kant notes that infants do not smile or cry tears until they are about three months old. He uses this as evidence that we need to first be able to form ideas—the idea of kindness or the idea of offense—in order to have these feelings (A 127).[11] In other ways Kant continually reiterates that feelings do not exclusively come from our animal nature, or even our human nature; they can also come from our rational, moral nature. Kant's discussion of feeling shows us that pleasure itself is not psychologically primary. Pleasure is an effect, and the causes of pleasure are susceptible to moral hierarchy: the greatest feelings come from the greatest causes.

Although it is perhaps necessary, I hesitate to discuss the relationship between feeling and desire because, for our purposes, the most important thing to say is that the capacity to feel pleasure and pain and the capacity to desire are, for Kant, two entirely different mental faculties. Nevertheless, misunderstanding Kant's theory of emotion (or failing to see it in the context of his theory of feeling) usually stems from focusing on Kant's theory of desire, specifically his theory of moral motivation, and from there by concluding that, for Kant, moral actions cannot stem from empirical motivations and that therefore emotion has no role to play in his moral theory. Kant's occasional attempt to explain moral feeling in terms of a feeling that comes *before*, instead of *after*, an action is not very helpful for charting the morally acceptable relationship between feeling and desire or the difference between moral feelings and immoral feelings.[12]

Kant's theory of moral motivation changed throughout his writings. What remains constant is the idea that we must do the right thing because we understand and appreciate that it is the right thing, not because doing the right thing is going to benefit us or please us in some way. In his *Lectures* Kant remarks that pure practical reason cannot in itself be sufficiently motivating.[13] Guyer believes that this conclusion is Kant's considered opinion on the topic, but I believe that Kant later realized that moral comprehension and moral feeling are two sides of the same coin.[14] In other words, the Kant of the *Lectures* holds that overdetermination is necessary, but the later Kant realized that overdetermination is inevitable, and hence redundant. If there was a significant development in Kant's theory of moral motivation in the time span between these two texts, we can at least agree that it was to realize that the necessary cooperation between reason and feeling entails more of an overlap between these two faculties than he previously realized.[15]

Sorensen taxonomizes Kant's theory of emotion primarily in terms of a distinction between feelings that are not necessarily related to desires and those that are. This classification may be confusing because it seems that whether or not a feeling is connected to desire is an incidental quality of it; feelings are better characterized by their causes, not by their effects, especially since Kant holds that we can choose whether or not we act on any of our desires. At base, feeling and desire are two different faculties. It is true that Kant highlights the fact that some feelings (like aesthetic feeling) are not connected to desire. This emphasis is because Kant is always skeptical of the possibility that our pleasure-seeking desires will sway our decision making, and he wants to create theoretical space for the notion of moral pleasure. Nevertheless, moral pleasure is like aesthetic pleasure because it is not pleasure-seeking, not because it seeks no action.

When establishing the role that feeling plays in performing our moral duties, Kant is careful to distance feeling from inclination or desire. Kant consistently speaks disparagingly about "inclinations" (*Neigungen*), making the point that our moral duties are not the same thing as our desires, no matter how rational—or rationalized—we can make the case for following them. He is attempting to change the way his readers think about feeling. Normally one considers feeling to be something merely passive, as a passive response to an experience that makes us have a certain desire (i.e., makes us want to do something). Nevertheless, some feelings, he points out, are not connected to desire at all. First, there is intellectual pleasure or displeasure that is entirely intellectual and not immediately connected to action; second, there are the feelings (satisfaction or dissatisfaction) that might follow from a consideration of our own actions (but do not precede those actions; DR 211). Overall, Kant is here trying to break the common assumption that feelings are necessarily connected to desires. Pace Sorensen, he is not trying to establish that some feelings *are* necessarily connected to desire.

Nevertheless, we would hope that all of our moral feelings would be connected to moral actions, and that would mean that they would have to be connected to desire. The best way to classify feelings on a Kantian model is the way that he does at the beginning of the section on feeling in the *Anthropology*, in terms of their particular cognitive causes. If a feeling follows from moral ideas, it is a moral feeling. Similarly, it is only this type of evaluation that can instruct us on whether or not to act on a given feeling.

No matter which theory of moral motivation we favor, it remains true that for Kant, moral feeling, specifically the feeling of respect, *comes from* our comprehension of the moral law.[16] Accordingly, my expansion of Kant's discussion of feeling should in no way be taken as assimilating Kant's moral theory into moral sense theory. Moral sense theory is an approach to moral theory that takes morality to be based on certain natural emotions. A moral sense theorist might say something like the following: we all naturally feel sympathy, or we all naturally take care of our family members, and this is what we call morality. Instead, it seems clear that Kant is right to say that morality is a consideration of what we should do, rather than what we in fact do and that, furthermore, our "natural" feelings are not always good. Instead, moral feelings, on Kant's account, follow from our rational comprehension of the moral law; they do not ground our comprehension of the moral law. Kant writes:

> By categorical imperatives certain actions are permitted or forbidden, that is, morally possible or impossible, while some of them or their opposites are morally necessary, that is, obligatory. For those actions, then, there arises the concept of duty, observance or transgression of which is indeed connected with pleasure or displeasure of a distinctive kind (moral feeling), although in practical laws of reason we take no account of these feelings (since they have nothing to do with the basis of practical laws but only with the subjective effect in the mind when our choice is determined by them, which can differ from one subject to another). (DR 221)

Accordingly, whether it is our rational comprehension of the moral law or the feeling of respect that follows from it that spurs us to act is not as important as the simple fact that legitimate moral feeling is always the effect of rational comprehension.

It is to a fuller illustration of the moral feeling of respect, as well as its connection to the breadth of moral feeling, that we now turn. Kant generally refers to respect for the moral law as the prototypical moral feeling, but there are others, perhaps even many other different types of moral feelings. We can say that Kant uses the term "moral feeling" in both a narrow and a wider sense: in the narrow sense it refers to the feeling of respect for the moral law, and in the wider sense it refers to all of the feelings that necessarily, in practice, follow from respect for the moral law (DV 464).[17] We will discuss each in turn, starting with respect for the moral law, and we will then broaden Kant's discussion so that it will be perhaps easier for the reader to relate to it.

Respect for the Moral Law

Respect is one of the most important Kantian concepts, in addition to reason and freedom. Not only do we have a duty to respect others, as well as ourselves (a point that will be discussed in the next section as well as the following chapters), Kant also describes our subjective experience of the moral law as one of awe-filled respect. Although some may doubt it, given Kant's reputation for exalting pure reason, respect is a bona fide physiological experience—a feeling. In fact, it can be a *very* emotional experience. The respect we feel for the moral law is the primary instance of this feeling and the other forms of respect follow from it. The full scope of moral feeling is even greater than those emotions that are variants of respect, but even the scope of respect for the moral law is wider than the term "respect for the moral law" connotes since there are multiple possible occasions for comprehending the moral law. In fact, we will see that the notion of respect merges with the other sense in which Kant uses the term "moral feeling," namely to refer to the physiological dimension of conscience.

One of the most famous quotes from Kant concerns the feeling we have when we think about the moral law:

> Two things fill the mind with ever new and increasing admiration and awe, the more often and steadily reflection is occupied with them: the starry heavens above me and the moral law within me. Neither of them need I seek and merely suspect as if shrouded in obscurity or rapture beyond my own horizon; I see them before me and connect them immediately with my existence. (CJ 161–162)

According to Kant, we know what is right and wrong when we think about it, and the feeling that this comprehension of moral duty creates is one of admiration and awe. It is very much like the feeling we have when we look at the amazing night sky, filled with billions of stars that are billions of miles away; we are captivated. Like the universe, morality is bigger than us. Of course, we can ignore morality, just as we can ignore the stars. Although it might be hard to imagine, a particularly selfish or frustrated individual might respond to the awesome night sky with discomfort or aggression.[18] Nevertheless, the night sky, like moral duty, is there, at the periphery, waiting to be seen for what it is.

In the case of the night sky and in other ways, Kant makes a connection between the feeling of respect for the moral law and the feeling of sublimity. As with the sublime, when confronted with the moral law, the subject feels both a loss of his own individual subjectivity and a strengthening of his rational power. Kant mentions that we feel a sense of the sublime in the face of the moral law in both the *Critique of Judgment* and in *Observations on the Feeling of the Beautiful and Sublime*, and we could possibly argue that the feeling of respect for the moral law is only an *instance* of the feeling of the sublime.[19] For example, Kant writes, contemplating a noble character is an occasion for feeling the sublime:

> [I]f the secret tongue of his heart speaks in this manner: "I must come to the aid of that man, for he suffers, not that he were perhaps my friend or companion, nor that I hold him amenable to repaying the good deed with gratitude later on. There is now no time to reason and delay with questions; he is man, and whatever befalls men, that also concerns me." Then his conduct sustains itself on the highest ground of benevolence in human nature, and is extremely sublime, because of its unchangeability as well as of the universality of its application. (O 65)[20]

Situations that cause this feeling might include those in which a person does the right thing in the face of horrible circumstances or risk of injury.[21] When we are exposed to moral greatness in any form, we have certain feelings. It can make us feel open and calm. We are likely to cry. In fact, the elicitation of tears is a common feature of many different types of moral feeling. We feel a sense of intense focus on the moral goal and awe. As with many emotions, the preoccupation verges on timelessness, and we feel as though we will never (or should never) leave that state. Depending on the immediacy of the situation, we might shake from being overpowered and have a rush of adrenaline.

A commitment to the moral law is a commitment to sacrifice oneself if necessary. Indeed, according to Kant, one must not deceive or disrespect others *even to save one's own life*. Therefore, proper comprehension of the moral law rightly inspires fear for our own survival but also a sense of being greater than slavishly serving the goal of survival. Instead, we feel a sense of our higher, moral purpose. Respect is a feeling of a worth that "thwarts my self-love" (G 401).

The feeling of freedom is closely related to the feeling of respect because moral autonomy is the expression of the highest form of freedom for Kant. The dynamic sublime is "an aesthetic judgment [in which] we consider nature as a might that has no dominance over us" (CJ §28). The feeling of respect for the moral law, which is akin to a mixture of fear and inclination or attraction, makes us feel that we are stronger than nature, specifically physiological determinism. We are free. The feeling of the sublime allows us to feel our independence from nature, an independence that keeps "the humanity in our person from being degraded" (CJ §28).[22] If we consider relegating the feeling of respect to merely an instance of the feeling of the sublime, we should also consider the possibility that autonomy and the moral law are paradigmatically sublime:

> Hence, if in judging nature aesthetically we call it sublime, we do so not because nature arouses fear, but because it calls forth our strength . . . to regard as small the objects of our natural concerns: property, health, and life, and because of this we regard nature's might (to which we are indeed subject in these natural concerns) as yet not having such dominance over us, as persons, that we should have to bow to it if our principles were at stake and we had to choose between upholding or abandoning them. (CJ §28)

Here we can see that Kant defines the sublime essentially in moral terms.[23]

Comparing and contrasting the feeling of respect to Kant's remarks on moral or religious fervor is instructive. We can think of examples of people becoming extremely emotional because of religious or moral ideas; Kant calls this enthusiasm (*Enthusiasmus*, more neutrally, and *Schwärmerei*, more pejoratively).[24] Enthusiasm can be problematic if it arises from false moral ideas or it if leads to imperfectly moral actions, but it cannot be distinguished because of its degree of feeling. In fact, Kant seems to think, as Sorensen points out, that we are naturally "enthusiastic" about morality.[25] We will address the way we should respond to moral feelings in the following chapters, but here we must note that the feeling of respect is a strong, physical feeling. Kant writes that instead of worrying that the pure presentation of the moral law would be unemotional and unmotivating, the truth is

> exactly the other way around. For once the senses no longer see anything before them, while yet the unmistakable and indelible idea of morality remains, one would sooner need to temper the momentum of an unbounded imagination so as to keep it from rising to the level of enthusiasm, than to seek to support these ideas with images and childish devices for fear that they would otherwise be powerless. (CJ 274)

We should not worry that Kant is grounding morality in emotion. The feeling of respect is inspired by comprehension of the moral law, not the other way around.[26] Kant writes:

> But though respect is a feeling, it is not one received through any influence but is self-wrought by a rational concept . . . What I recognize as a law for myself I recognize with respect, which means merely the

> consciousness of the submission of my will to a law without the intervention of other influences on my mind. (G 400)[27]

Kant calls the feeling of respect "self-wrought," but this does not mean that the bond between the cause of moral comprehension and the effect of the feeling of respect is any less natural. In fact, Kant's ethics relies on this degree of natural moral sensibility.[28] We naturally feel respect when we conceive of the moral law, even though the moral law is not a product of our animal nature—an instinctual response to external stimuli—but of reason.

Is it paradoxical that we have a natural feeling of our independence from nature? Sokoloff argues, along these lines, that respect is neither completely sensible nor completely intelligible; it is both and neither at the same time. It is a transient that eludes both poles of the binary Western opposition between reason and feeling.[29] For this reason Sokoloff maintains that Kantian respect is a paradox and reading it as such can help us suspend the tendency of "cognitive domination."[30] Nevertheless, it seems to me that Sokoloff's conclusion itself fails to truly challenge dichotomous thinking. If the feeling of respect is truly successful in eluding the reason/feeling dichotomy, why is it a paradox? If reason can itself be naturally (even instinctually) linked to feelings, then the feeling of respect does not constitute a "paradox" but an occasion for us to realize that it is only a deficiently understood brand of cognition that threatens domination in the first place. There is no reason, philosophically or emotionally, to shy away from the importance of reason in our lives; we are, after all, rational animals.

In the *Groundwork*, Kant introduces the feeling of respect for the moral law with great hesitation, defending his appeal to "an obscure feeling" with the retort that *this* feeling is different in kind from other feelings (G 401n).[31] Kant is perhaps worried that pure reason cannot itself motivate us to do anything and that if it makes use of anything other than pure reason, reason will not be pure. He is perhaps falling prey to the easy confusion between instinctual desire and feeling that occurs when the distinction between impulse and reason is made too simplistically. He seems worried here that all feelings are directed by our natural, selfish desire for happiness, although that is clearly not the case for respect.[32] By the time of writing the *Metaphysics of Morals* Kant has a much richer notion of moral feeling, but even his remarks in the *Groundwork* open up the door for other feelings that are based in reason.[33]

Respect for Others and Love for Others

Because humans are those beings who can feel respect for the moral law, respect for the moral law leads to respect for self and respect for all humanity. We might characterize this move from respect for the moral law to respect for humanity as the first step in the enlargement of the feeling of respect, but the next step, as we shall see, is more like an explosion into a universe of moral feeling. When translated into real relationships, the abstract admiration we owe to individuals because they approximate the ideal of reason quickly becomes a variety of more heated and complex feelings that thereby demand even more effort to evaluate.

Human beings are *rational* animals, capable of discerning right from wrong for themselves and directing their own lives. We therefore have the duty to respect others. Similarly, human beings are rational *animals*, who are born helpless, remain dependent, create new families and relationships, and require social and economic cooperation. We therefore have the duty to promote the happiness of others. To ourselves we owe the duty to strive for self-perfection because, while we are animals who naturally look after our physical well-being, we are morally called on to improve ourselves. While moral feelings accompany all of these pursuits, we shall here focus on the feelings that accompany our duties to others.[34]

Love and respect are the feelings that relate to our duties to others (DV 448). Both love and respect involve attention paid to another and care.[35] Furthermore, they each involve a number of more particular sympathetic feelings, which we will discuss next. Kant describes love and respect together—they are necessarily linked—as a mix of attraction and repulsion (DV 449). This metaphor risks confusing the reader, I think, if it is taken too literally.[36] It might be more helpful to describe love and respect as a mix of lowering others (or putting oneself above others) and raising others up (or putting others above oneself). Of course, we should never think that one person has more value than another, but one person might have more power than another. Usually promoting happiness involves giving someone something, whether a service, goods, money, and so on. When we are beneficent with another, for example, the gift establishes the giver in a position of power and the receiver in a position of dependency. In order to counter this inequality, respect exalts the other, by arguing that the other deserves the gift or that one offers it as a tribute, and so on.[37] Our relationships with others require a bit of a balancing act so as to approximate true equality. We can see that love and respect are similar feelings, but they differ with respect to one's position of power in the relationship.

We might say that with respect the person respected has power because he or she deserves certain things (e.g., the right to think for herself). Although love also describes a feeling of giving and admiration, the giving is freely undertaken and the receiver is thereby put into some form of an obligation of repayment. It is thereby wrong to allow someone to love you without reciprocation, and this is the reason that we hide unreciprocated love. Nevertheless, we do not hide our respect for someone nor do we expect the recipient of respect to repay the gift. With love, we expect reciprocation, but with respect there is some reason to justify the gift.

Is it right to think of generosity in terms of repayment, as it appears that Kant does? Kant's point is not that it is right, but inevitable. The needy often do not want to be the recipients of charity; charity is more pleasurable and fulfilling for the giver rather than the receiver. (Aristotle similarly notes that the best life involves some amount of superfluous wealth so that the wealth-holder can take part in the virtue of charity.) Nevertheless, sometimes we need the help of others, no matter how unpleasant it is. We must realize that people need each other's help, but when giving it, we must be sensitive to the feelings of the receiver. Kant helps the giver navigate the feelings of the receiver by suggesting that the poor deserve charity and that it is thereby not even meritorious for the wealthy to give it to them because economic inequality is most often caused by unjust governmental institutions in the first place (DV 453). Therefore, the wealthy are merely giving that which rightfully belongs to the poor.[38]

We must also always remember that we have the duty to promote the happiness of others, not to promote the perfection of others. Generosity should never go without respect. Respect in its original meaning of awe-filled amazement, or even sublime subsumption of self to other, is required. We cannot always understand the purposes of another, nor should we necessarily always try to—except to help her further them. If we hope to understand her goals in order to evaluate them, we are not thereby *helping* her or fulfilling our moral duties. As rational beings, we communicate with and learn from each other, but such educational "help" should not be construed as generosity. Sometimes it is quite difficult to figure out how to help others, as it might be difficult to ascertain why they are making the choices they are making. Of course, we are not *barred* from questioning the motives of others, nor are we barred from refusing to help at any point, especially if we suspect that we might be contributing to immoral activity. Nevertheless, *not helping* should not be construed as helping. In some cases, it is easy to figure out how to help others, as with giving them food when they are hungry, for example (not a lot of self-reflection can take place on an empty stomach); other cases are more difficult.

When someone is beneficent with us, we are obligated to be grateful. Kant writes that we are obligated to have gratitude because, pragmatically speaking, if gift-receivers did not show gratitude it would weaken the desire to be beneficent in the giver. Of course, we are morally obligated to be beneficent whether or not we ourselves receive anything from it (like the pleasure we might take in seeing the other person happy or reveling in their gratitude), and so anonymous generosity is sometimes the best. Still, pragmatically and anthropologically speaking, we humans are more likely to do good if we take some pleasure in it, and so finding, taking, and giving that pleasure is also morally required (DV 454–455).

Amazingly, in the case of gratitude, we actually have a duty just to *feel* grateful, according to Kant, and this feeling need not be connected with any action. This makes sense. We might be obliged to say "thank you" when someone is generous with us, but, according to Kant, the real duty is not just to say "thank you" but to actually *feel* thankful (DV 454–455). Kant's frequent objection that we cannot be obligated to have a feeling since we are not free in our feelings does not come out here. We can clearly do certain things to help ourselves to feel grateful if we do not, but it is quite interesting here to see the extent to which virtue for Kant involves *feeling* the right thing at the right time. Furthermore, in this case, a feeling is morally necessary because of the relationship it has with the feelings of others.

SYMPATHY

Loving and respecting people, in real life, requires that we sympathetically take part in their lives.[39] Physiologically speaking, sympathy is an interesting concept because it can involve various possible feelings. It might involve sadness or anger or joy. (We must guard against the prejudice that some feelings, like anger, for example, are necessarily bad.) Like love, it often involves concern, which can be described as a focusing of attention and a quickening of the pulse. In a sense, it merely means taking the fate of the other as your own, so if something good happens to another person, you too are happy, and if something bad happens to another person you are sad or angry. Often though, sympathy is not merely reactive, and it involves worrying about what one can do to help.

Kant writes that we have a "duty to cultivate the compassionate natural (aesthetic) feelings in us" (DV 457). It is here that Kant famously remarks that we should not

> avoid the places where the poor who lack the most basic necessities are to be found but rather to seek them out, and not to shun sickrooms or debtors' prisons and so forth in order to avoid sharing painful feelings one may not be able to resist. (DV 457)[40]

Kant is explaining the fact that we naturally have certain sympathetic feelings—that fact is something over which we have no control. Yet we do have control over whether or not we expose ourselves to the stimuli that trigger those feelings. Someone who bars oneself from such stimuli might be described as actually uprooting the natural feelings of sympathy in himself; he might become so unused to experiencing sympathy that he might not even completely feel it when he does, accidentally, become exposed to an unfortunate situation. On the other hand, if we do routinely expose ourselves to situations involving people with whom we will sympathize, we will have more and more occasion to exercise our duty to help people.[41]

As many feelings are involved with sympathy, more are involved in repressing it. I venture that most people avoid the contemporary version of "sickrooms and debtors' prisons" as much as they can by staying in certain parts of town and sufficiently guarded institutions—even our laws protect us from those "unfit" for proper society. And, when we might happen to cross paths with an "undesirable," most people have become very adept at simply ignoring him, with such strategies as avoiding eye contact. We have even become adept at ignoring or demonizing "moralizers," like Kant, who might call on us to sympathize. Nevertheless, ignoring moral duty involves ignoring or repressing moral feeling. Uncovering (or further creating) moral denial and repression then becomes an emotional endeavor all its own.

It is of course possible, as readers of Kant are aware, that we might allow ourselves to be ruled by the *feeling* of sympathy and not by proper comprehension of our moral duties. Kant examines this mistake in the *Groundwork*, and his discussion of sympathy there has been chiefly responsible for the poor reputation of his philosophy of emotion. Since any mention of Kant's theory of emotion most commonly calls to mind Kant's criticism of the moral motivation of sympathy, it is best that we devote some time to elucidating his position.

In the *Groundwork,* Kant uses an example of the difference between someone who is charitable because she is naturally sympathetic and one who is charitable even though she is not naturally very sympathetic. He uses this example in order to show that the most important thing in morality is one's moral principles and one's understanding of what is right and wrong.[42] Although the moral feelings we have been discussing here are those that follow from a proper understanding of our moral duties, not all feelings are so clear-sighted. Some feelings—Kant calls them affects—follow from thoughts that have not been well understood or evaluated. These feelings may, in fact, develop into or from genuine moral feelings, but they also may not. Similarly, the way we react to feelings is important. We must understand the thoughts and perceptions from which they originate and evaluate their legitimacy. A good understanding of our moral duties must be the basis of any choice, never a mere feeling.

Herman gives a perhaps humorous example to help explain this point. We might, she writes, feel sympathy for someone carrying a heavy load. We might naturally want to help him. We might even think to ourselves, "It is my duty to help others." Still, if this person is a hooded figure carrying a heavy painting out the back door of an art museum at night, we should probably not help him. The feelings of sympathy should not make the final decision.[43]

It is true that some versions of sympathy seem to be insufficiently rational and hence insufficiently moral. It is therefore correct to insist, as does Kant, that moral feelings must be based on proper moral comprehension, or if they do seem to rise up quickly, they must prove themselves to be in line with the demands of morality on further reflection. In the "Preface" to the "Doctrine of Virtue," Kant calls following moral feeling instead of moral reason "blindness." We are reminded of his remark in the first *Critique* that intuitions without concepts are blind. It is not that any feeling, taken by itself, is bad, wrong, or unnecessary; they are simply blind. A blind person can find his way around, but he does so either by following (or feeling) something else or by mere chance. The feeling of sympathy, if experienced without any explicit moral deliberation, might lead us immediately to the right course of action, but it does so without certainty.

Of course, moral duty often calls for quick action, and we might not always have time to evaluate our feelings fully before acting on them. Nevertheless, the insight that feelings can spring from rational comprehension gives us reason to act in these times of crisis, and not, as the critic of reason might suppose, to hold off. Mistakes are of course possible, since we always have limited information, but we are nevertheless called to do the best we can. Moral deliberation, feeling, and reflection must be a continual process. Indeed, this process is perhaps a good description of the virtuous disposition.

In helping to explain the difference between sympathy in a moral sense and an insufficiently moral sense, we can look to Kant's distinction, in the *Anthropology*, between being sensitive and being sentimental. He suggests that virtue is premised on the former and threatened by the latter (A 235–236). The sensitive person is aware of other people's feelings; he understands their causes, and he is patient with them. The sensitive person is quiet and sad when another suffers a loss; he is pleased when another is proud. Furthermore, this sensitivity allows him to see the ways in which he might be able to help. A sentimental person allows the other's feeling to become amplified. In the case of sympathy, this amplification is often felt to be an intrusion. In the case of one's own emotional experiences, the sentimental person allows his emotions to call the shots. He most likely even enjoys being driven by the strength of feeling instead of by rational (considerate) reflection.

Unfortunately, the *Groundwork* example mentioned above has often been misread to imply that it is better to act without moral feeling than with moral feeling. It is most accurate to say that we should act *with* feeling but not *from* feeling.[44] A case of moral action without moral feeling may be better for seeing an example of moral principle, but it is not psychologically or pragmatically better. (It might not even be physiologically possible.) As we have already seen, Kant writes in the "Doctrine of Virtue" that we have a duty to cultivate virtue and that virtue is a psychological disposition to do the right thing. Moral feelings, like sympathy, play a large role in this psychological disposition. Nevertheless—Kant's point from the *Groundwork* still stands—they do not play the only or determining role. Kant is not a moral sense theorist, that is, while he does hold that we have natural moral feelings, he does not ground morality in them. Indeed, we also have natural immoral feelings, like envy and malice. Moral judgment tells us which of these feelings we should cultivate and which ones we should reexamine.

Cultivating Moral Feeling

Kant's stance on the duty to have moral feelings is not always easy to understand. We do not have direct control over our feelings because they are the *effect* of cognitions. Nevertheless, moral feelings help us to accomplish moral actions. Therefore, insofar as we have the duty to do moral actions, we have an indirect duty to feel moral feelings. There are a variety of ways we can feel more moral feelings, starting with the most direct route of carrying out (or exposing ourselves to) more moral actions and extending to the least direct route of attempting to experience other sorts of unselfish feelings.

We have the moral duty to cultivate moral feelings.[45] "Moral feeling," in a narrow sense, refers to respect for the moral law, as we have already seen, but in a broader sense it refers to a general sensitivity to the demands of morality, and, in the broadest sense, it refers to the feelings, like properly grounded sympathy and love, that these demands require. All of these virtuous feelings follow from proper comprehension of our moral duties.

If the feelings naturally follow from the moral law, why do we have a duty to cultivate them? In one respect, we don't. In explaining the cultivation of virtue, in the "Doctrine of Virtue" Kant argues that some aspects of morality cannot be described as virtues, per se, for they are simply natural:[46]

> There are certain [natural] moral endowments such that anyone lacking them could have no duty to acquire them—they are *moral feeling*, *conscience*, *love* of one's neighbor, and *respect* for oneself (*self-esteem*). There is no obligation to have these because they lie at the basis of morality, as *subjective* conditions of receptiveness to the concept of duty, not as objective conditions of morality. All of them are natural predispositions of the mind (*praedispositio*) for being affected by concepts of duty, antecedent predispositions on the side of *feeling*. To have these predispositions cannot be considered a duty, rather, every human being has them, and it is by virtue of them that he can be put under obligation— Consciousness of them is not of empirical origin; it can, instead, only follow from consciousness of a moral law, as the effect this has on the mind. (DV 399)

It makes sense to think that if these feelings are simply natural reactions, then they are out of our control, and we thereby could not possibly be obligated to have them. Similarly, Kant is here saying that the feeling of respect for the moral law is a part of what it means to respect, that is, be obligated by, it. To argue that we are obligated to be obligated is nonsensical.[47] Therefore, Kant seems to be arguing that we do not have an *obligation* to have moral feelings "because they lie at basis of morality" and hence at the basis of obligation itself (DV 399).[48]

In order to not get confused, we need to remember what Kant means by feeling. Feelings are the physiological effect that certain representations have. Kant is right to say that, in one sense, there is nothing we can do to make ourselves feel certain things. Feelings follow automatically, in this case naturally, from certain thoughts;

there is nothing we can do to establish these original connections. Nevertheless, in another sense, there is plenty we can do to make ourselves have certain feelings, namely we can think certain thoughts or engage in certain behaviors. Therefore, as we have seen, we can contemplate the moral law and our moral duties in order to strengthen our moral feelings. An earnest commitment to doing good is the first step to discovering our duties and therefore doing them. Kant argues, for example, that it is nonsensical to prescribe a duty to have a conscience: "To be under obligation to have a conscience would be tantamount to having a duty to recognize duties" (DV 400). Nevertheless, this is only because Kant believes that we already do, naturally and necessarily, have a conscience. We can choose whether or not we listen to it. "Cultivate" is the best word because we cannot plant the feelings, as it were, but we control how well they grow.

Kant clearly holds that we do have the duty to cultivate moral feelings, even though they are natural, because he defines virtue as strength in overcoming those inclinations that oppose the moral law, and moral feelings are the main source of that strength. Immoral inclinations might be an active opposition to morality or might be mere weakness, such as the lack of courage. While different immoral feelings will require different strategies for their overcoming, we gain the positive strength of virtue both by contemplating the moral law and by acting virtuously, thereby practicing virtue (DV 397). So, for example, it does not make sense to say that we have a duty to first *feel* benevolently; we have a duty to understand the need for benevolence—it follows from the fact that humans are finite, dependent creatures—and act beneficently. It is likely that benevolent feelings will follow:

> *Beneficence* is a duty. If someone practices it often and succeeds in realizing his beneficent intention, he eventually comes actually to love the person he has helped. So the saying "you ought to love your neighbor as yourself" does not mean that you ought immediately (first) to love him and (afterwards) by means of this love do good to him. It means, rather, do good to your fellow human beings, and your beneficence will produce love of them in you (as an aptitude of the inclination to be beneficent in general). (DV 402)

Aside from deliberating about our moral duties and strengthening moral feelings by practicing the fulfillment of them, there might also be some other, indirect ways we can cultivate moral feelings. In the third *Critique*, Kant writes that the feeling of beauty is morally instructive because it teaches us to appreciate unselfish pleasure. Aesthetic feeling, like moral feeling, is an example of a feeling (pleasure) that is occasioned by intellectual activity. Kant discusses the means by which the natural aesthetic response can help one develop a sense of shared humanity and thereby, indirectly, mutual respect. Furthermore, aesthetic feeling is based on disinterested observation; it is disconnected from the promise of physical pleasure. Kant calls the beautiful the "symbol of the morally good" because both require pure intelligibility, the experience of freedom, and the unity of the theoretical and practical powers, as well as the idea of the supersensible substratum of nature that allows for its harmony with freedom (CJ §59).[49]

Lest one disparage aesthetic judgment for being merely a *symbol* of moral judgment, we must note that aesthetic judgments exemplify many of the key features of moral judgments. For example, aesthetic taste is morally instructive because it teaches us to have a purely intellectual liking (CJ §IX). Similarly, our presumption that our aesthetic judgments should be universal might directly lead to some moral judgments, like the moral necessity to preserve nature, for example.

Iris Murdoch is critical of Kant's philosophy, but her own characterization of "the Good" demonstrates similarities with Kantian moral theory. She also describes the good with an analogy to aesthetic beauty. As with Kant's aesthetics, Murdoch takes the beauty of nature as the purest form of beauty, suggesting that we sometimes intentionally "give attention to nature in order to clear our minds of selfish care."[50]

Murdoch discusses the morally educative content of many great artworks:

> The death of Patroclus, the death of Cordelia, the death of Petya Rostov. All is vanity. The only thing that is of real importance is the ability to see it all clearly and to respond to it justly which is inseparable from virtue.[51]

It is true that art often deals with death and human frailty. It does seem likely that some works of art could direct us toward moral feelings, although we should note that art is no more prone to the presentation of moral exemplars than the opposite. It is also possible that people are susceptible merely to imitating art, which would undermine its capacity for truly instilling moral feeling or commitment.

Science is like art for Murdoch in that it also teaches us to respect the search for truth over our own self-interests.[52] It does make sense that intellectual virtues, like curiosity and integrity, would not only follow from moral duties but could lead to our sensitivity to them. Nevertheless, moral truth is not identical to scientific truth, and there is some risk in confusing the two. Moral truths are discovered by reason and not by observation; still, acting on them does require some degree of scientific insight.

We are now concerned with identifying the ways we can *make* ourselves have certain feelings. Yet perhaps we have taken a wrong turn since, as we already noted, we naturally already do have moral feelings. In order to complete our discussion of the methods we can use to cultivate moral feelings, we need to take a step back and reconsider the true breadth of moral feelings. Geiger suggests that the scope of moral feelings might be as large as the range of our duties, and that is what we were beginning to see above.[53] If we see and accept the vast array of moral feelings that we already have, cultivating them becomes the task of more fully expressing them.

MORE AND MORE MORAL FEELINGS

One leaves a nice restaurant fully satiated. Walking down the street, a homeless woman approaches. She asks for nothing except for the leftovers from the meal. *Here they are.*

One feels sad: *So many people are poor when we are so lucky.* One feels regret: *Was there something else I could have done to help her?* Other moral feelings crowd in: happiness at having done a good deed; conscience demands to know whether or not there was a moment of hesitation (after all, those were really good leftovers). When we start to look, we see more and more moral feelings.

In Kant's texts, the term "moral feeling" most often refers specifically to the feelings one has when thinking about the moral law or in response to a specific moral action. It might seem that thus far we have only named a small group of moral feelings and that these feelings are rather esoteric, but, in fact, moral feelings pervade our lives. We can see this already from the fact that in particular cases there will need to be different feelings that respond to the specifics of the situation. Kant remarks that there will need to be an "aesthetic of morals," which would be a complete elaboration of all of the many feelings that go along with moral decision making. Kant does not undertake such a task, as it likely cannot be completed. He merely mentions in passing that there is in fact a great variety of feeling (e.g., disgust, horror, etc.) that accompanies the subjective presentation of moral obligation (DV 406). In addition, Kant remarks that "there is no human being so depraved as not to feel an opposition to breaking [the moral law] and an abhorrence of himself" (DV 379). We must therefore add the moral evaluations of our past behaviors onto the catalog of moral feelings. I will give but a few examples to show how varied the class of moral feelings truly is. (In doing so, I will also have to cover the nature of our moral relationship to animals.)

I remember the first time I cried from happiness; I was around six years old.[54] I remember that I had previously thought that it is very strange that people cry out of happiness because I could plainly judge that people cry out of sadness. It was at an event tailored to trigger the emotions: a movie in a movie theatre—a children's animated classic of my generation. At the end, the main character (an animated mouse) is reunited with his family, and I, along with most of the people in the theatre, most naturally, cried. This example is no different from the tears that are elicited by most "happy endings"—to which I am perhaps more susceptible than others—and it merely sticks out in my memory because it was my first experience with "happy tears." Nevertheless, many of the moviegoers experienced sympathetic happiness for the larger-than-life animated mouse. There seems to be little reason not to call this very common experience of emotion moral feeling.

Perhaps one might object that this is merely an affect, prereflective, and not grounded in a proper comprehension of the moral law. There is some truth to this claim. I myself was only six and had only the most basic understanding of morality. Additionally, the presentation was only fiction, and the character was not even a person but a talking mouse. This expression of feeling seems to cross the line into sentimentality that we charted earlier. Perhaps the viewer should be more critical of the intentions of the moviemakers than I was. She should ask, for example, whether it is right to feel sympathy for a mouse. My suspicion is that the majority of the moviegoers would be willing not only to separate a mouse from its family but kill it in other circumstances—but, of course, that other mouse would bear little resemblance to a human being.

The objection succeeds in showing us the extent to which the feeling of sympathetic happiness in this situation was perhaps prereflective. Nevertheless, there are also ways in which it did in fact follow from a proper comprehension of the moral law and could translate into better comprehension thereof. We can see that the main character's happiness depended on being reunited with his family, and we moviegoers had no selfish reason for caring about his happiness.[55] We sympathized with him only because we perceived him to be (sufficiently similar to) another human being, and we too care about our families.

There are ways that acting on this feeling might, if prereflective, lead us astray. The children who watch that film might decide that they love mice dressed in clothes, and they might goad their parents into buying movie paraphernalia. Kant's worry that affects might lead us into actions that we have not fully thought through also applies to moral feelings. Even comprehension of the moral law might lead, as we have seen, to a kind of moral fervor or enthusiasm that, as Kant writes about affects, destroys its own purposes. Nevertheless, the feeling of joy that a little mouse has been reunited with his family has a number of positive effects, especially if we focus on this feeling itself and try to exclude other affects that are mixed with it. This experience, most likely, made me more sensitive to the plight of refugees and my own connection with my family. Like the many other moral feelings people experience, it made me, in a small way, into a better person. Nevertheless, as with any feeling, its connection to action cannot be taken for granted.

As we look at the next example, we will want to keep in mind the meaning of tears as well as the possible responses one might have to the purportedly moral feeling.[56] Recently, I was working in a community garden. A man there was volunteering to help his mother with the gardening, and I directed him to hoe up a thistle bush so that his mother might plant in its place. Soon enough the man realized that he had in fact hoed into a rabbit's nest. The mother rabbit fled and a small bunny was left hiding on the ground. I found the bunny, and—deciding with limited information about the best course of action—put the baby bunny into a box with some of the fur and leaves that had made up the nest and put it into my car to take home to nurse to maturity. I judged that action to involve the lesser of two evils, namely the mother losing her baby and the mother abandoning the baby and the baby dying.[57] (Later, upon instruction from my veterinarian, I returned it to the garden because I was told that it would likely live on its own and would likely not respond well to being in captivity.)

After putting the bunny in the box, I suspected that the man believed that he had killed a rabbit because he saw loose fur, and I could also tell that he was upset. I endeavored to ease his mind some, and I told him that he had not killed a rabbit, that they make their nests from loose fur, and that he should not feel guilty because he had not hurt the rabbit on purpose. It was just an accident. Nevertheless, he told me that he had cried. Poor man—I become teary now recounting it.[58] That this man cried reflects favorably, I think, on his character—as Kant lectured, "We can judge the heart of a man by his treatment of animals" (L 240).

Are the tears of this man an example of moral feeling? My guess is that he felt sorry that he had unnecessarily hurt an animal. I do not know him, but I assume that he eats meat—perhaps hypocritically—without crying, and so I assume that he cried because he was confronted with the case of the unnecessary suffering of an animal at his own hand.[59] While we do not have a duty to promote the happiness of animals in the same way that we have a duty to promote the happiness of people, we have a duty not to cause them needless suffering (DV 443). Kant writes that cruelty to animals

> dulls [one's] shared feeling of their suffering and so weakens and gradually uproots a natural predisposition that is very serviceable to morality in one's relations with other men. (DV 443)

The man in this example was feeling this "natural predisposition" to a "shared feeling of [animal] suffering" to which Kant refers.

Is this an example of a moral feeling? Or are the tears an example of a prereflective affect that might happen to be morally useful? Just as in the previous example, the response that follows from this feeling could, in fact, be bad. The man may have repressed his sadness and it may have turned into a feeling of embarrassed weakness and hatred for all that is weak. (Of course, he did not do that.) Even if he did not at all repress the feeling, he might have still not fully understood it and integrated it into his life. The real question might be whether or not he would have shown any further immediate concern for the rabbit itself. Perhaps, had I not been there, his confusion and embarrassment would have prevented him from figuring out what his duties in that immediate situation were. Kant writes that affects often stymie their own purposes; in either of the two previous examples, the tears may have led to embarrassment thereby eschewing any moral deliberation. The example here is, like the previous example, of a relatively prereflective moral feeling; that is to say, the moral thoughts that caused it were likely buried among other thoughts or not fully formed. Nevertheless, in either case, with further reflection they could be discovered and worked out; our moral commitment—feeling, comprehension, and action—would thereby be strengthened.

Although it might be paradoxical, the strongest experience of moral feeling might be triggered by confrontation with human weakness and moral failings. We experience feelings of moral horror, moral fear, moral tragedy—really all tragedy elicits moral feelings. It is the great disappointment in our inability to do as much good as we would like that both overwhelms us and keeps us convinced of the moral fire burning inside us. Similarly, it is our propensity to criticize our surroundings that keeps us attached to the moral realm. Art often expresses social criticism, as well as the accompanying meditation on human frailty and the fragile and simple touches of human nature and goodness. In effect, any time we believe that the world should not be the way that it is, we are having (at least the beginning of) a moral feeling.

Lest I lead the reader to think that merely feeling something makes it true, we must further discuss the proper and necessary response to moral feeling. Often the proper response is to act on moral feeling, but never simply because we *feel* something to be the case. Legitimate moral feeling must be based in reason.

INFALLIBLE FEELING?

If Kant places such emphasis on moral feelings and respect for the moral law, does he believe that someone can be wrong about what the moral law is? Perhaps Kant is open to the criticism that he makes moral knowledge seem too easily gained. In fact, he is often criticized for reducing the entire moral sphere to one simple rule, the categorical imperative. This strikes some readers as both overly simplistic (ignoring the complexity of the variety of possible situations in which one might find oneself) as well as overly rigid (again ignoring the complexity of the variety of possibly situations in which one might find oneself). I have dealt with the criticism that Kant ignores the necessity of contextualism elsewhere, but here we need to give more detail about what moral decision making looks like for Kant, in order to show that he does not imagine that for any particular moral situation the heavens open up and the moral law tells us what the right thing to do is.[60] Nor are we are overcome with an infallible feeling that only needs be followed. In fact, the opposite is the case: it is often difficult to make good decisions.

Although we might reserve the term "moral feeling" for feelings that are either prompted by the moral law or seek to evaluate our past behaviors, Kant does not think that all feelings that are experienced as moral feelings are necessarily legitimate. Kant criticizes moral enthusiasm when it is not in line with considered moral principles (DV 408–409). Perhaps we can find an example of mistaken moral feeling in Kant's discussion of masturbation and nonprocreative sex. Although Kant perhaps admits that we might sometimes need to violate a prohibition on non-procreative sex "in order to prevent a still greater violation" (such as adultery), he writes that "the thought of it...stirs up aversion" and this "unnatural vice" makes everyone feel ashamed (DV 425).[61] Here he is assuming that everyone is naturally disgusted by masturbation. This is probably the case of taking what we would all now probably think is a nonmoral or confused feeling for a bona fide moral feeling. How do we tell the difference?

The next chapter is devoted to evaluating our feelings morally, but here we should make clear that the feeling of respect is not only a product of reason but also conducive to reason, and that it need not be "trusted" or blindly followed any more than reason itself needs to be "trusted." "In fact," Kant writes:

> [N]o moral principle is based, as people sometimes suppose, on any feeling whatsoever. Any such principle is really an obscurely thought metaphysics that is inherent in every human being because of his rational predisposition. (DV 376)

In other words, while feelings often come from our sense of right and wrong, there is no guarantee that that sense is well worked out and rationally grounded. Feelings and moral deliberation are wrapped up together, both in varying degrees of awareness and reflection. Our task in perfecting ourselves is to become more aware of our feelings and the thoughts that ground them—the "obscurely thought metaphysics"—so that we can evaluate their moral cogency.[62]

Still, the criticism at hand is that the emphasis Kant places on the feeling of respect makes it seems like grasping the moral law is easy and that it is impossible to be morally mistaken. I am not sure how to respond to that criticism. Studying Kant over the years has been helpful for me to better grasp the content of moral duty, and I do think that Kant is right about the moral law and our two specific duties of virtue. Nevertheless, it also seems to be the case that in everyday moral decision making we might be wrong about what we think we are morally obligated to do. We might put too much weight on promoting our own nonmoral self-perfection instead of promoting the happiness of others (or vice versa); we might believe that merely sending money to a charity is a good way to promote the happiness of others; we might conclude that cutting off contact with people who make immoral choices or are insufficiently virtuous is necessary for our own moral self-perfection. These decisions might be a moral mistake, and we will hopefully, over time, realize the ways that we can improve. Nevertheless, I am hard-pressed to come up with an example of ways that we might be wrong about the bare content of moral duty itself. Still, to be on the safe side, we must assume that abstract moral reasoning could be mistaken, and we should take steps to check, for example, for impure motives or unintended consequences.

In addition, we must keep in mind that much of the content of morality is left up to individual moral deliberation. That fact does not mean that there is no right answer; instead, it means that, in the majority of cases, no one has figured it out yet. Kant tells us what the moral law is. Although there are multiple formulations of the categorical imperative, he takes them all to express the same idea, and whenever he states *the* moral law, he always gives the first formulation of the categorical imperative: "[A]ct upon a maxim that can also hold as a universal law" (DV 225). Nevertheless, the moral law is abstract; it is devoid of all empirical information. It must be mixed with information about specific contexts in order to yield any determination of what to do at all. Practical reason means thinking for ourselves, and we are all fallible.

Furthermore, contrary to popular opinion, ethics is, according to Kant, exactly that sphere of governing actions that does not give us any advice about what to do. Ethics tells us what our necessary ends must be—our own perfection and the happiness of others—but it does not at all tell us how to promote them. Ethics is the opposite of the sphere of juridical right because the laws of justice (i.e., the laws it would be just for a government to have) tell us what we can and cannot do but they do not tell us what we can and cannot want. Nobody but you can tell you what to want. Moral decision making is our most intimate realm of self and personhood; it is a task we cannot alienate.

Because it is only the ends of action that are specified by ethics, not the actions themselves, these duties are "wide" and there is some "playroom" for us in deciding how we are to enact them (DV 390). By "playroom" Kant does not mean to refer to the *extent* to which we pursue our own perfection and the happiness of others in contrast with amoral pursuits. (I will address the question of "how much" in chapter 7.) He means instead that we must decide for ourselves *how* we are to perfect ourselves and make others happy. In perfecting ourselves, we must decide where and how we are going to seek education, for example, or practical psychological advice about emotions. We must decide who will be the beneficiary of our generosity and how best to give without making the other feel embarrassed and indebted. We must discern, even though we are not responsible for the perfection of others, when we should contribute to another person's efforts and when we should not.[63] We also have to find ways to remind ourselves to do all of these things so that we do not neglect our positive duties entirely.[64]

Feelings play a role in reminding us that we have moral duties and in helping us discover the best ways to discharge them. As the discussion above suggests, we already do feel a number of moral feelings. The task then is to reflect and act on them. In the ideal case, moral feelings follow from moral consciousness. Nevertheless, in real life, moral feelings might be uncoupled from moral consciousness. I happen to think that feelings often involve particularly intense thought content, but, perhaps because the person is uncertain in her thinking, the feelings rather than the thoughts take center stage.

If feelings are the subjective, physiological effects of objective thoughts and perceptions, and thoughts and perceptions can vary in their degrees of explicit awareness, then it stands to reason that we might sometimes be more aware of our feelings than we are of their causes. Similarly, feelings, because they are physical, might be harder to avoid than thoughts.[65]

In Kant's discussion of conscience we see his recognition of the layers of consciousness and the difficulty we have in coming clean with ourselves, so to speak:

> Every human being has a conscience and finds himself observed, threatened, and in general, kept in awe (respect coupled with fear) by an internal judge; and this authority watching over the law in him is not something that he himself (voluntarily) makes, but something incorporated in his being. It follows him like his shadow when he plans to escape. He can indeed stun himself or put himself to sleep by pleasure and distractions, but he cannot help coming to himself or waking up from time to time; and when he does, he hears at once its fearful voice. He can at most, in extreme depravity, bring himself to heed it no longer, but he still cannot help hearing it. (DV 438)[66]

Here we have the case of feelings or symptoms that have been uncoupled from their cognitive causes because the latter are to some degree repressed. Certain behaviors might then be like psychic traces that we are forced to follow back to the repressed content, if we are morally oriented, lest we continue to repress them, creating ever-removed symptoms. Skill in psychologically diagnosing ourselves and handling the complexities of feeling and thought, then, is not only psychologically but morally necessary.

It might sometimes seem that we are letting feeling lead the way, but Kant remarks that that would be "to reverse the natural order of cognitive powers" (A §4). Strictly speaking, it is not possible to let feeling lead the way, as the worry that we are taking feeling to be infallible would suggest. Even if we are not aware of the thoughts that have caused a particular feeling, the feeling was still the product of some act of cognition. Similarly, thoughts are a part of a longer process of rational and moral deliberation. If the feeling is itself a sign of anything, it is that we should not prematurely abandon that deliberative process.

CHEERFUL VIRTUE

In cultivating moral feelings, the goal is a virtuous character, with the experience of moral feeling as a well-integrated part of life.[67] Virtue, for Kant, is a cheerful form of strength or courage that follows from the intellectual orientation of continual moral commitment. The better we understand Kant's notion of virtue, the more the differences between him and Aristotle become minimized. Nevertheless, Kant holds that virtue will always be somewhat of a struggle. Even the most perfect person is no angel, and so, referring back to the moral law and asking ourselves what we should do, "virtue is always in progress and yet always starts from the beginning" (DV 409).

Annas criticizes modern ethical theory, of which Kant is her main example, for holding that virtue must always correct emotions, and she argues that Ancient ethics is superior because it accepts that sometimes feelings can lead *independently* to the right result.[68] Annas argues against Sidgwick's conclusion that we are not responsible for our feelings and so they do not belong to ethical theory, holding instead that we are responsible for our feelings, although not in the moment.[69] She gives the example of working to break a bad habit. Annas's most well-known criticism of Kant is that, having assumed that all emotions are vicious, he mistook virtue for mere continence, that is, mere control over emotion and not the presence of virtuous feelings.[70]

Pace Annas, we have seen both that Kant does think that we have some control over our feelings and that he has a relatively robust theory of moral feeling. In addition to moral feelings, Kant also discusses other types of emotions, what he calls affects and passions. We will turn to these in chapter 6. Kant clearly holds that some emotions—and I take "emotion" to be the most general term encompassing all of these various topics—are more prone to virtue while others are more prone to vice. Similarly, some emotions are more a product of conscious reflection and others are more a product of suppressed or confused thinking (perhaps responding quickly to external stimuli). The problem with prereflective feelings for Kant (what he calls "affects") is not that they are necessarily vicious but that they are prereflective, as we will see more fully in chapter 6. Neither Kant nor Aristotle holds that all feelings are virtuous.

Kant acknowledges the importance of natural moral feelings as well as their proper cultivation, but he defines moral worth in terms of correct understanding. Annas's characterization of Aristotle's theory seems to include the idea that it would be desirable to try to use feeling *exclusively* as a guide, attempting to cut it off from all rational reflection. It seems to me that if someone wishes to argue that it is possible to do the right thing *without knowing what the right thing to do is*, the burden of proof is on her to show how this is possible. Similarly, where would moral feelings come from if they were completely *independent* from all forms of cognition?[71] In arguing that Aristotle recognizes that good habits must be based in proper moral understanding, as Annas begins to, the sharp contrast between Kant and Aristotle on this score seems to fade.[72]

We must note that Kant characterizes virtue as a kind of fortitude against the continued resistance from *inclination*; inclination (*Neigung*) is a very different notion than feeling (*Gefühl*). When we confuse "inclination," which does connote immorality for Kant, and "feeling," which does not, we are similarly tempted to oppose feeling and reason, but, as we have seen, that opposition is impossible by Kant's definition of feeling.[73]

Still, Kant calls virtue fortitude—later, he calls virtue a "strength of resolution" (DV 390). Virtue stands in opposition to "that which opposes the moral disposition *within us*" (DV 380).[74] In my mind, the jury is still out over whether it is better and more accurate to think of immoral inclinations as relatively intractable or as easily overcome. Perhaps it is better to conceive of virtue as a struggle in order to emphasize that we are constantly pursuing virtue, unlike Aristotle's assumption that citizens of an ideal state are already virtuous, resting on their laurels, as it were. Kant writes:

> But virtue is not to be defined and valued merely as a aptitude and . . . a long-standing habit of morally good actions acquired by practice. For unless this aptitude results from considered, firm, and continually purified principles, then, like any other mechanism of technically practical reason, it is neither armed for all situations nor adequately secured against the changes that new temptations could bring about. (DV 6:383–384)

Perhaps it is the case that Kant is considering virtue in the real world of corrupt society, not in an ideal state. Even so, virtue is an achievement of an individual for Kant—indeed, the highest achievement—and not of a society. Even if both thinkers hold that virtue follows from knowledge, Aristotle, as a nominalist, holds that moral knowledge comes from experience. Kant holds that the world is too immoral for that to be possible.

Kant does not think that virtue can be described as a habit, that is, as unthinking behaviors, because morality must always first and foremost be based on the understanding of moral principles. Nevertheless, in the process of trying to perfect oneself we may sometimes need to help ourselves do the right thing by making it easier and more automatic. There is nothing wrong with forming good habits; there is only a problem with losing touch with their grounding moral principles.

Say, for example, that I know that I should help people when they need help, but I really do not know whom to help or when to help. In order to spare myself the mental work of constantly having to think about who and when to help, I just settle into the habit of always giving money to a homeless man I see on one particular corner on the side of the road on Saturdays. Assuming that my generosity is actually helpful, there is nothing wrong with this habit; on the contrary, it is good. Nevertheless, if it fully eclipses further thinking about my duties to help people or the evaluation of the effectiveness of this action, it is clearly problematic. In fact, if I stop traveling down this route on Saturdays, I might cease fulfilling my duties to help other people entirely simply because I have failed to make it a *mental habit* to stay in touch with the theoretical content of my duties.

One problem with thinking that moral habits or moral feelings can lead to good behaviors independently of moral evaluation is that this claim fails to recognize the psychological complexity of human cognition and reinforces a dichotomy between thinking and feeling. On the Kantian model human thinking is complex and we must coordinate many different levels of experience at one time. Indeed, understanding and harmonizing one's own variety of different thoughts is itself difficult, let alone the number of feelings that arise from them and elsewhere. Practical moral judgment involves the task of weaving together these multiple strands of experience. We have seen that virtue requires the psychological skill of sensitivity and of working *with* our feelings as well as the feelings of others.

Cheerfulness, not continence, is the ideal of Kantian virtue.[75] Kant argues that we will not be successful in doing good if we merely forbid ourselves to follow our immoral inclinations. Instead we must find a way to reduce the strength of these immoral inclinations by confronting them on their own turf, as it were. If reason simply tries to overpower feeling, it will lose because feelings and immoral inclinations are too strong. The duty to "rule oneself" goes beyond "forbidding [oneself] to let [oneself] be governed by [one's] feelings and inclinations" (DV 408). We must cultivate virtuous feelings. Kant criticizes "monkish ascetics," which aim merely to dominate and repress sensual inclinations:[76]

> The cynic's purism and the hermit's mortification of the flesh, without social good-living, are distorted interpretations of virtue and do not make virtue attractive; rather being forsaken by the Graces, they can make no claim on humanity. (A §88)[77]

It is remarkable, and yet often overlooked, that Kant sides with the Epicureans over the Stoics regarding moral pleasure (DV 6:485).[78] He argues that if we do not find pleasure in moral behavior we will shirk our duties:

> The rules for practicing virtue (*exercitiorum virtutis*) aim at a frame of mind that is both valiant and cheerful in fulfilling its duties (*animus strenuous et hilaris*)....what is not done with pleasure but merely as compulsory service has no inner worth for one who attends to his duty in this way and such service is not loved by him; instead he shirks as much as possible occasions for practicing virtue. (DV 484)

It is not enough to merely put up with misfortune, as Stoic training aims; one must enjoy life. Furthermore, not having conflicting motivations itself makes us happy. This kind of happiness is an expression of what Kant portrays as the highest good in life: happiness in proportion to virtue.

A proper understanding of virtue yields the realization that we must constantly remain open and sensitive to feelings—being sensitive but not sentimental. In addition, virtue itself is described in emotional terms, both as courageous and cheerful. After all, the virtuous person is in touch with the feelings and relationships that give life its true value. Accordingly, we should not be led to think that moral feelings are solely selfless and altruistic. Far from it: enjoying our relationships is one of the most important aspects of virtue.[79]

I conclude with the example of the Kantian ideal of moral friendship to show again that, for Kant, the virtuous life is not only cheerful and pleasant but also full of a variety of feelings. True friendship, for Kant, is one that is not based on a passing appreciation of someone's pleasant company. Rather it has weathered the test of time and is a moral expression of mutual respect and aide. Kant calls this "moral friendship," the "most intimate union of love and respect" (DV

6:469). Kant admits that true friendship is "unattainable in practice," but to strive for it is a "duty set by reason" nonetheless (DV 469). Friendship, for Kant, requires the "equal balance" of feeling and duty; one must be very careful to strive for this balance lest one err on the side of coldness or on the side of disrespect. Friendship is manifested in helping one's friend, and this help is an expression of "inner heartfelt benevolence" (DV 471). Friendship also involves two people sharing their feelings with each other:

> moral friendship (as distinguished from friendship based on feeling) [*zum Unterschiede von der ästhetischen*] is the complete confidence of two persons in revealing their secret judgments and feelings [*Empfindungen*] to each other. (DV 471)[80]

Nevertheless, even the most virtuous person probably still needs to "control" her feelings occasionally, and it is to this topic of emotional evaluation that we now turn.

5

Emotional Universalism and Emotional Egalitarianism

In order for us to get a sense of what is involved in evaluating emotions, we need to have a deeper discussion of moral theory. In chapter 3 we saw that emotions need to be understood and evaluated—they often even involve their own self-evaluation. Then, in chapter 4 we saw that many emotions are moral in nature but still need to be evaluated and acted on. This chapter will focus on the specific processes by which we should evaluate our emotional experiences, and the next two chapters will focus on the phenomenological dimension of vicious and virtuous characters and qualities in general.

Kantian moral theory is often explained in terms of the different formulations of the categorical imperative, or the moral law, especially the first two.[1] This chapter will be roughly divided between these two moral rules, explaining them and their relevance to our emotional lives.[2]

The first formulation is often called the "Universal Law formulation" and it runs thus: "Act only according to that the maxim whereby you can at the same time will that it should become a universal law" (G 421). Following Arendt, I liken Kant's universalism to the notions of personal and political transparency. Kant places a great deal of emphasis on transparency, which begins with self-transparency or conscientiousness, but this fact is often missed because attention is diverted by the too fine of a point Kant puts on the injunction not to lie. In terms of conscientiously evaluating our emotions, we find that feelings and ideas can be repressed either because they are selfish or because they are not fully understood. Emotional denial can also take the form of a lack of moral courage. Kant's emotional universalism entails a process of both moral and psychological therapy, moving from the unconscious and possibly selfish to the well understood and inclusive perspective.

The second formulation is often called the "respect for rational being" formulation, and it runs thus: "Act in such a way that you treat humanity, whether in your own person or in the person of another, always at the same time as an end and never simply as a means" (G 429). We should understand the moral notion of respect not just as a passive leaving other people alone but as active consideration. Respect is an integral part of developing close relationships as well as navigating all varieties of conflict. Indeed, we will see that respecting oneself and respecting others are interrelated. In turn, we must evaluate our emotions in terms of whether or not they respect the self and others.

MORAL UNIVERSALISM

As is well known, the first formulation of the categorical imperative (CI) states that one must never act in such a way that one could not also will that the maxim of the action be a universal law (G 402). This idea puts us on the lookout for any intentions that we would not find acceptable in someone else in the same circumstances. We normally call the actions that follow from such intentions "unprincipled." Kant explains this maxim test both through the notion of a contradiction in conception (as with the lying example) and a contradiction in willing (as with his example of refusing to be charitable).[3] In the first case, there are some intentions that directly contradict themselves. For example, lying undermines its own vehicle of rational communication; cutting in line contradicts the concept of a line; cheating breaks the rules according to which one hopes to win. One might object that some rules need to be broken or that some lies are necessary, and to respond to such objections we will most likely need to go beyond the notion of a mere contradiction.[4] The second type of contradiction, a contradiction in willing, begins to merge with the second formulation of the CI, referring to respect for all rational beings, since it adds into the thought experiment the notion that the will itself has a certain nature. For example, there might be no pure contradiction in my willing to refuse to offer help to others, but given that I myself am a being that will need (and has needed) help, the performative act of such refusal is contradictory. You cannot know, for example, that refusing to be charitable or even murder are not universalizable unless you know that the being doing the moral thinking is a human being, that is, one dependent on external help, with a living body, and so on.[5] It might be the case that in order to make complete sense of the notion of universalizability you must join it with a notion of human dignity, which is separately explained in the second formulation, but

Kant himself fleshes out, as it were, the notion of a universal law, with his idea of a universal law with a law of nature. Kant writes that when taking the effects of our actions into account, moral thought requires that we think in terms of a harmonious, or natural system, so he writes that the Universal Law formulation of the CI can also be expressed thus: "Act as though the maxim of your action were to become through your will a universal law of nature" (G 421). Perhaps this is a better way of expressing it, as is the Kingdom of Ends formulation, which also considers the whole of the community, since it shows us that Kant is not unconcerned about the effects of action, just unconcerned with them in the immediate sense.[6]

Critics of Kant have argued that the CI yields incorrect dictates or that it does not give us enough information about what we should do. These two criticisms relate to its negative and positive functions in turn. In the first case, (contradiction in conception) the CI generates perfect duties, that is, actions that we should never do.[7] In the second case (contradiction in willing), it specifies general necessary maxims but not specific actions (imperfect duties). The second criticism—that it is too vague—is easier to dismiss since it is true that our imperfect duties cannot be fully specified ahead of time. Each of us must use our own moral judgment to figure out when, where, and how we should promote the happiness of others and our own natural self-perfection.[8] The second criticism, that we do not have the moral requirements that the CI concludes, is more difficult to address.

The most popular objection to the CI, that lying is not always wrong, was first voiced while Kant was still alive by Benjamin Constant. Constant proposed the situation where a person helps to hide another person from a would-be murderer. If the murderer asks the person where his sought-after victim hides, he should not tell him the truth. Kant responds to this scenario in his "On the Supposed Right to Lie for Philanthropic Purposes," developing the notion that blame accrues from immoral actions not from moral ones; hence the truth-teller is not to blame for the murder. While this limitation to culpability makes for a good moral theory, we might strengthen his defense a bit with further explanation. It should be noted that this criticism cuts to the heart of Kantian deontology, which holds that actions can be wrong in themselves without reference to their consequences. It is *always* wrong to lie because it disrespects the humanity of another. All people deserve respect regardless of their past or future deeds.[9] Although it might be improper to refer here to the murder's motivations, it is likely that the above-mentioned murderer has become murderous from having not been sufficiently respected. It is wrong to treat the would-be murderer as a mere monster.[10] Nevertheless, one should not *help* him commit murder. The guardian should do everything in his power to protect the victim, and he need not offer up information that can be used for harm. Nevertheless, he cannot lie; he might say "I refuse to help you, sir, because what you are about to do is wrong," but he should not lie. Again, although it might be improper to gesture to positive or negative consequences, being treated like a monster and being lied to is not likely to calm the potential murderer down. Of course, the morally acceptable course of action requires far more bravery than merely lying.[11]

The requirement of universalizability might also be expressed as the requirement to think about morality and value it highest. The maxim of an action expresses the principle on which we act, and oftentimes the same action can be represented by different principles. For example, if I win a raffle and claim my prize, I might be expressing the principle of fairly following the rules or I might be acting on the principle of taking the prize no matter what. Even if I do not do anything wrong, my intentions can be immoral. For Kant, morality must be understood from an internal perspective: we judge actions based on the principles, or maxims, they embody. In every choice that we make, we are expressing some principle or another, that is, some hierachization of values—and one of those subjective values is morality. In chapter 7, we will see that morality requires that one have a moral orientation in life overall. In everything we do we must be concerned with our moral duties, so as both not to transgress them and to actively fulfill them. If we act only for our own purposes, whatever they may be, without checking into their consequences or whether or not they square with respecting ourselves and others, then there is some degree of moral failure with our choices whether or not we do anything wrong. Instead we must think about our choices from an expanded, moral perspective; our intentions must always be at least partly moral.

The notion of moral and emotional universalism highlights the common ground that all people share, even if it might take some patience to discover it. We can see this notion of universal human nature more fully in Habermas's explanation of universalism: universalism holds that moral justifications must be, in principle, acceptable to all rational beings.[12] By that, he means to exclude reasons that cannot, in principle, be acceptable to all rational beings because they are based on something essentially particular to one person or group, such as faith or other personal, idiosyncratic, or cultural practices.[13] A stronger reading of this notion of "acceptability" posits that all rational beings would find the same thing acceptable, since they are all rational, and there is presumably a most rational answer to any question, if it can be found—this is the ideal of not just compromise but rational consensus.

Habermas's assumption—that any person from anywhere could potentially speak and agree with anyone else—may strike us as odd and counterintuitive—even offensive—especially given the popularity of cultural relativism and our current culture wars. James Rachels offers a nice response to the popularity of moral relativism.[14] He gives the example of infanticide as supposed evidence that different cultures have radically different moral codes. He argues that if we probe this practice we will find that it is not so radically different after all. Nowhere do parents routinely kill their babies wantonly or gratuitously; it is only done out of necessity and sorrow, after seeking out adoption possibilities. Assuming commonalities allows us to see that we have a duty as humans to reach out to others and connect, no matter how different they may appear. Similarly, we must stand up to immorality and injustice wherever it occurs. It is heartening and liberating to focus on the common ground of emotion and reason that all people share. To dismiss the possibility of understanding because another person comes from another country, speaks another language, or has different political views is not only morally disrespectful but also morally dangerous.

Even still, the ideal of rational communication should not be seen as excluding the possibility of learning from others. Arendt interprets Kantian universalism in the spirit of pluralism, arguing that universality is achieved by taking on many different perspectives.[15] Arendt's formulation of the CI expresses Kant's emphasis on the importance of overcoming selfishness.[16] Arendt draws a connection between the Universal Law formulation of the CI and what she calls the "Transcendental Principle of Publicness" (TPP) from *Perpetual Peace* (PP). Therein Kant states:

> All actions relating to the right of other men are unjust if their maxim is not consistent with publicity...[for a] maxim which I cannot divulge publicly without defeating my own purpose must be kept secret if it is to succeed; and, if I cannot publicly avow it without inevitably exciting general opposition to my project...the opposition which can be foreseen a priori is due only to the injustice with which the maxim threatens everyone. (PP 129–130)

Kant believes that people have an innate sense of justice. Publicly communicating an unjust maxim would arouse opposition and would therefore cause others to prevent the action. Arendt quotes Kant from *The Strife of the Faculties*:

> Why has a ruler never dared openly to declare that he recognizes absolutely no right of the people opposed to him? The reason is that such a public declaration would rouse all of his subjects against him; although, as docile sheep, led by a benevolent and sensible master, well-fed and powerfully protected, they would have nothing wanting in their welfare for which to lament. (SF 145)

Any of us might be dictatorial and selfish in our refusal to consider others in our decision making.

The TPP expresses the same sentiment as the CI, and it helps us to gain a more intuitive grasp of Kantian universalism. Still, the TPP might seem closer to Habermas's description of universalism than to Kant's first formulation of the CI since one might keep something a secret because it is not, in principle, acceptable to all rational beings, while, on one possible reading of the Universal Law formulation, we can universalize something that might not be explicitly acceptable to all and vice versa. The TPP may be seen as more stringent than the Universal Law formulation since it gives the power of dissent to others, and the Universal Law formulation allows the individual actor to decide herself, in conducting a thought experiment, on behalf of others. One would hope that the moral decision-maker would decide in the same way that the other person would if the latter were given a chance to speak for herself, but such is not necessarily the case, and so we might conclude that the TPP formulation actually does a better job of respecting autonomy than the Universal Law formulation of the CI because it requires real, not imagined consent.[17]

Universalism is best understood as requiring impartiality and the overcoming of selfish motives. Both universalist formulations of the CI (the TPP[18] and the Universal Law formulation) require a transition from the judgment of acceptability made by one person to the judgment of acceptability made by all. Arendt argues that the "bad man," for Kant, is the one who "makes an exception for himself."[19] That is a good way to put it. Universalism also implies an expanded perspective of one's goals and motivations: they must be evaluated not merely as they relate to oneself but as they relate to everyone.

Let us not forget how simple at least one dimension of universalism is. My four-year-old son, without, I think, any pedantry has recently grasped it. At the grocery store, I told him I had to bring our cart back to the front of the store. He responded, "Or else you will get punished." I was puzzled. "I won't get punished," I retorted. Then I thought about it more: "Sometimes people do things not because they're afraid that they'll get punished if they don't but just because they want to do the right thing. I can't leave the shopping cart in the parking lot or else cars wouldn't be able to get around." He butted in—"What if everyone did that?" This simple question, "What if everyone did that?" is so useful. It might not express the absolute value of humanity—we'll cover that in the next section—but it does express the basic moral point of view (as Peter Singer puts it).[20] Sure, I would like to leave my shopping cart in the middle of the parking lot. I also like to buy plastic things, take plane trips, and drive in gasoline-powered cars. Nevertheless, there's a fundamental difference between my personal inclinations and impersonal moral reasoning because, after all, "What if everyone did that?"[21]

EMOTIONAL UNIVERSALISM

To come to see one's motivations from an honest, expanded perspective is also the goal one takes up in understanding and evaluating one's emotional experience. Hence, emotional self-improvement involves something like emotional universalism. Both the process of achieving moral universalism and the process of achieving emotional universalism encourage us to take on other people's points of view and to look at ourselves from the outside in, as happens in many forms of therapy. Moral universalism is itself a form of therapy: in striving for universalism we achieve a better understanding of our standpoint, and, in some cases, overcome it.

Kant's insistence on truthfulness needs to be understood in the psychological context of achieving self-knowledge. In his discussion of moral character in the *Anthropology*, he argues that truthfulness is a necessary prerequisite to character:

> Briefly, as the highest maxim, uninhibited internal truthfulness toward oneself, as well as in the behavior toward everyone else, is the only proof of a person's consciousness of having character. (A 295)[22]

What does it mean to have "uninhibited internal truthfulness toward oneself"? This sounds intense, not to mention naïve after we have accepted the insights of Freud. Kant's own belief that we can never be fully aware of whether or not we have a purely good will seems to suggest that total internal truthfulness is impossible. Nevertheless, the duty of self-knowledge requires that we strive for it. If we can never fully be aware of our intentions, then the effort to be truthful with oneself, or not to lie to oneself, needs to be constant. We need not be paranoid, but an intention of which we are unaware cannot be morally evaluated, and so we must be psychologically vigilant. Lying to oneself is even more morally corrosive than lying to others because correct motivation, which is premised on self-understanding, is necessary for morality. Furthermore, lying to oneself threatens the possibility of communication just as lying to others does. The goal of communication is community, and a community of rational beings is achieved through rational transparency; no communion can be reached if that which has been shared and understood is false.[23]

The first step of evaluating our emotions in terms of universalism is to understand them. Emotions are complicated, and we often do not know the reason that we are feeling the way that we are. We each have different strategies for figuring out what we are feeling. Whenever I feel particularly overcome by emotion, I find it handy to make a list of the different emotions I am feeling and the different possible emotions I might be feeling. I try to make it as exhaustive as possible so that I know I am not missing anything (and that there is a definite limit to the turmoil). Doing that helps me to determine which emotions are the strongest, that is, most important to address. Of course, sometimes some other internal force of repression stands in the way of our making sense of our feelings.

In the context of our discussion of emotion, it is safe to assume that the desire to keep an emotion secret could be related to its selfish intentions or a lack of moral courage. Emotions themselves can harbor selfishness, and selfishness, as we have seen, resists disclosure. Therefore, selfishness is usually marked by defensiveness or anger. To be vigilant in scanning one's emotions for selfishness—or at least to acknowledge that one should overcome selfishness when one finds it and should not resist finding it—involves little more than believing that selfishness is bad. While some economists proclaim that selfishness is good, most normal people would not do so. There might be a possible willingness to keep quiet about selfishness, but that does not mean that it is acceptable. My guess is that some people live by principles of—at least limited—selfishness. They believe, most likely inconsistently, in self-preservation. When they form friendships or get married, they believe that they then live to help themselves, their friends, spouses, and children. When people have not reflected on their moral principles, they often act for a mix of moral and immoral reasons. Reflecting on these principles, and subjecting them to something like the requirement of publicness, will help them overcome selfishness.[24] Seeking out possible moral claims others might make on oneself, by not avoiding poor houses and sick rooms, for example, or any other moral feeling we might have, shows us that we often try to get away with selfishness by means of avoidance.

Emotions are not specifically prone to selfishness just because they are feelings and self-oriented (such an assumption would involve a confusion between "selfish" and belonging to one's self), but they may be specifically prone to selfishness in that they are prereflective and, hence, making an exception for ourselves might be repressed and hence manifest itself in physical form. We already considered the way that many people adhere to a whatever-my-emotions-say-goes approach to emotionality; that approach may be motivated by defensiveness about being wrong and the desire to reduce cognitive dissonance. Emotions such as angry resolve and vindictive bitterness, that is, emotions that actively resist change and the calming force of reflection, may be the most likely vehicle for hidden selfishness. Perhaps the most common type of selfishness embedded in emotions is simply our not wanting to admit that we are wrong. The reasons behind our emotions may be repressed, as Averill suggests, for the very same reason that the despot will not announce publicly that he holds his subjects in sheer contempt: he does not want to be discovered.[25]

In addition to the possibility of selfishness, we must not overlook the role that fear plays in the lies we tell to ourselves. There are many possible reasons that we might be afraid of our emotions; thankfully, they are usually quite banal. Mostly, we are afraid of our own weakness. Negative emotions usually relate to our limited powers and needy nature; hence acknowledging them requires vulnerability. Positive emotions also require the courage to acknowledge vulnerability. In order to love and be happy, one must be willing to acknowledge and express need and be willing to connect oneself with another. One needs as much courage to be happy as one does to be sad. Repressing an emotion out of the desire not to feel or appear weak or needy is disrespectful to oneself, and so it is perhaps best covered in the next section.

Since human life is characterized by interconnectedness and emotion, emotions are themselves universal in many respects, and the moral dimension of emotion just as often requires that one accept and navigate emotions as it does that one overcome certain selfish emotions. The phrase "emotional universalism" may be misleading because it sounds as though everyone should have the same emotions.

There is reason not to adopt the term in the fact that Kant faults feeling for not being universal:

> The capacity for having pleasure or displeasure in a representation is called feeling because both of them involve what is merely subjective in the relation of our representation and contain no relation at all to an object for possible cognition of it (or even cognition of our condition). While even sensations, apart from the quality they have (of, e.g., red, sweet, and so forth) because of the nature of the subject, are still referred to an object as elements in our cognition of it, pleasure and displeasure (in what is red or sweet) express nothing at all in the object but simply a relation to the subject.[26]

Sweetness may not be a quality of a peach, but of our taste buds, just as respect is not a quality of the moral law, but there is still an objective, law-like connection between some objective experiences and certain subjective feelings or emotions. We might even say that a peach has the quality of being able to cause a certain taste when paired with human taste buds. The fact that the feeling is "in us" need not mean that it is disconnected from the object. Adam Smith argues that proper emotions are those that would be had by a detached observer.[27] This suggests that there is much that is common among emotional responses and that these commonalities ought to be seen as normative. Nevertheless, Smith's account is backwards: it is not the commonality that grounds morality, but morality that grounds the affirmation of the commonality. Also, we should not think that a "detached" observer would be an unemotional observer; instead, the detached observer would be like the subjects of the despot, offended when they discover the contempt he holds for his subjects.

The position that emotions are subjective, private experiences itself thwarts the call to universalism. Kant warns against the desire to "play the spy upon one's self," which "is to reverse the natural order of the cognitive powers" (A §4). Kant is referring to the tendency we see all too frequently of holding one's own idiosyncrasies to be precious simply because they are one's own. We can imagine someone who is protective of her emotions, someone who insists on her "right" to have them, since she cannot be mistaken about the fact that she is having the experience. Although feelings are subjective experiences, just as thoughts are, they, like empirical and cognitive experience, are imbued with intersubjective content and refer to objective states of affairs. In many ways, another person may be able to understand our feelings better than we do. For the most part, emotions are natural reactions to certain experiences. There is a vast array of possible situations and events, but, for the most part, all people understand the law-like connection between them and emotions. "Emotional universalism," then, refers to coming to an understanding of the connections between one's emotion and the universal laws that connect it to its causes. This understanding entails making the emotion transparent and grasping its cognitive content. If we have a "right to our feelings," it is only in the sense that we have a "right" to think freely, but all too often such a "right" is understood as license not to think at all.

Emotional universalism, then, should be seen as gaining greater perspective on one's emotions, not another person's external perspective, but a perspective others might have if they were in the same situation. In other words, we should try to imagine, not a detached but, a very affected spectator. Differing perspectives may bring to light a number of things: ways that we are limiting our experience of emotions, an understanding of the reasons that we have the emotions that we do, of the unconscious purposes our emotional habits serve, facts or feelings that our emotions imply, ideas in responding to our emotions, and so on. We cannot overemphasize the fact that, in many cases, taking a universal perspective of one's emotions entails becoming *more emotional*. A friend of mine related a story to me of her adolescent years, in which she was very depressed, nearly suicidal. In trying to comfort her, her half-sister said to her: "You are depressed because you were sexually abused by your brother." Although it would be obvious to any outside observer that sexual abuse would have horrendous psychological consequences for the victim, this revelation did not strike my friend as plausible. Instead, I suppose, she believed that the sexual abuse had little emotional effect on her. Later she, of course, realized that she had been in denial. All of us can most likely look back over the course of our life and find similar disavowed causes of our feelings. Sometimes a false account of causality—a lie—is created. For example, someone might believe that he or she has a reason for doing something—like canceling a meeting because of snow, for example—but the real reason is an emotion that the person does not want to acknowledge, like a fear of hearing what she suspects the other person will say. Such a lie that is told both to ourselves and to the other person is not only morally corrosive, it is psychologically corrosive as well. In addition, such a lie is usually quite transparent—Freud remarks that the unconscious of one person is usually able to pick up unconscious messages from another person.[28]

Emotions are apparently simple. It seems as though it would go without saying that sexual abuse, especially abuse that is not properly redressed, would make one angry, sad, frustrated, despondent, disgusted, ashamed, even suicidal. There is surely some kind of law-like fixity between these experiences and these feelings, and, yet, we humans seem to be bad at accepting the connection between thoughts and feelings and making good on it. Sadly, our emotional ineptness is often a social ineptness and, hence, a moral ineptness. Responding to the emotion means that we must respond to the events themselves, which can be so difficult that we deny the emotion. Much is at stake in many of our disavowed emotions and so courage is required to achieve emotional universalism; perhaps this emotional courage is the paradigmatic example of moral courage.

Critics of Kant might object that moral universalism and emotional universalism cannot be correlates because Kantian moral universalism, they say, requires that we overcome all particularity, which includes all emotion.[29] This criticism is based on confusion about what it means to take on a universal perspective. Moral universalism requires that our action be, in principle, acceptable to all rational/emotional people, not that all of our actions become uniform. As we have seen, there is no reason to think that "rational" means unemotional or that the majority of our emotions cannot be perfectly rational. Nevertheless, it is necessary to formulate the precise meaning of this "rational acceptability" in the case of emotion, especially since emotional self-improvement has often been misunderstood and taken to mean rational repression (as we will see in chapter 8). Consider the following example of evaluating whether or not an emotion is universalizable: Suppose I resist making time to visit and comfort a grieving friend. The emotions at work here may be a sort of anxiety, guilt, or a disingenuous arrogance and defensiveness about the importance of whatever it is I happen to be doing instead. Or, I might just "forget" about him. The goal of emotional universalism would be to understand the maxim that occasions the emotions and to morally evaluate them. This maxim is immoral because people need comfort in certain situations and my refusal to provide it makes an exception of myself: I cannot deny that people sometimes need comfort, nor that I will sometimes need it, but instead I want to opt out of being the one who must provide it, hoping that someone else will do the work for me. My maxim would be that everyone should provide comfort for grieving friends, except for *me*. Or perhaps I want my friend to like and support me, but I do not want to provide support for my friend. Or perhaps I think that I only want to be friends with happy people and I would prefer to live in a world in which sadness is just forgotten and people are not very deeply connected to each other. Again, such a maxim would not be universalizable since humans need to be deeply connected to each other for their own psychological health. Sadness at death and other times is good, or, at least, better than repression. In this case, I will need to judge the emotions that prompt this immoral maxim as defective, and I will need to judge some other emotions, such as compassion as well as fear and sadness at facing death, as important and requiring expression. In this case it seems that behavioral therapy (the behavioral therapy involved in facing one's fears so as to become more comfortable in certain situations) would also be necessary.

In this situation universalism is not blind to particularity. It does not matter that someone else, who is not friends with my friend, does not have the duty to comfort him. What does matter is that anyone can see that I, being the particular person that I am and in the particular situation that I am in, have a duty to behave thus. The right thing to do is to comfort my grieving friend.

There might be further reason to think that universalism conflicts with emotionality in that emotions can be idiosyncratic and a product of very particular facts about oneself, such as the particular personalities of one's parents. Let us say, for example, that I am rather neurotic about cleanliness and vacuum as often as I can. Is it okay that I am this way and others are not? Emotional universalism prompts us to understand our motivations and make sure that they are morally acceptable. Idiosyncrasy itself is not immoral, nor are any specific, individual desires; they are a problem only when they trump moral concerns. In other words, when we gain a more universal perspective on our principles and beliefs, we may decide to give them up or we might instead be strengthened in them. Emotions are a feature of individual experience and upbringing, it is true, but so are beliefs, convictions, and principles. Universalism does not mean that all individual experiences must be traded in for some kind of universal experience, whatever that could possibly mean. Our particular experience must be evaluated from a universal perspective, but the universal perspective remains a view of our very individual and particular life.

Emotional universalism requires acceptance of the fact and demands of emotionality, just as moral universalism requires acceptance of the fact and demands of morality. There is an easy transition to be made from becoming more aware of one's own emotional needs and the moral requirement of respect, which is our next topic: self-denial often takes a moral toll on others. In other words, people often deny the needs of others because they deny their own needs; they also deny harm caused to others because they deny harm caused to themselves.[30] Recognizing our own emotions teaches us something about emotions (although a proper theory of emotion also plays a role), and accepting our own emotionality promotes a feeling of equality. We become more emotionally literate in general and more comfortable with the fact that emotions are a part of humanity, as well as more sensitive to them. Doing so fosters moral courage.

MORAL AND EMOTIONAL EGALITARIANISM

Although, as we have seen, universalism requires that we recognize the equality between ourselves and others, the second formulation of the CI is better known for expressing this sentiment: act so as to always treat people (rational beings) as ends in themselves, never as merely a means (G 429). All of the formulations of the CI are meant to express the same idea, and with the TPP, we see that universality is related to rational acceptability. Universality is not only the form of reason, it is also the guarantor of inclusion. We must make sure that our actions are, in principle, acceptable to all and compatible with the actions of others, because all people count. All people count because humanity itself has inherent value—this is the content of the second formulation of the moral law.[31]

Respect can be misconstrued as leaving other people alone.[32] Kant's notion of respect is double-sided: it includes leaving others alone to let them direct their own lives but it also includes the duty to help them achieve happiness, however they might understand it. Respecting other people requires that we grant them a right to govern themselves as much as we are able to govern ourselves. Here, more explicitly, we see the interrelation of self- and other- respect. Our devotion to rational self-evaluation prompts us to realize that others need that same internal space to figure things out for themselves. Furthermore, respecting others requires that we understand and engage them emotionally. Respect is a feature of relationships, and relationships are emotional.

As we have seen, the notion of selfishness plays a major role in Kant's thought. Selfishness, for Kant, entails taking up the immoral maxim, privileging inclination over the moral law; instead, (as we will see in chapter 7) all self-worth and self-respect must be premised on respect for morality. To value the fulfillment of one's inclinations over the moral law means that one is willing to trample another person's (or one's own) dignity in order to fulfill a personal goal. Disrespect might take the form of lying, coercion, or simply a failure to communicate with another person and gain consent. Selfishness can be understood as valuing one's own goals too highly or as a failure to empathize and recognize the equal worth of other people's goals. Selfishness can also result from a lack of skills, those that are informally referred to as "social skills." Such skill is required for recognizing a human person as a human person in the first place. Without this recognition it is possible to live in an artificially depopulated moral world, caring about a few people perhaps, but ignoring many others. When others do impinge on consciousness, they are seen as obstacles, not as people. Insensitivity to the demands of morality is then a kind of mental self-centeredness.

Kant argues that we have the duty to promote our own self-perfection and the happiness of others, *not* the happiness of ourselves and the perfection of others. This distinction between promoting the happiness of others and promoting the perfection of others offers an interesting parallel here. We might be working very hard for others—indeed, devoting our entire lives to them—and still be failing to respect them. Kant argues that we have a duty to promote the welfare of others and a negative duty to promote their moral well-being. In other words, we must refrain from corrupting people but need not be their moral teachers:

> For the perfection of another human being, as a person, consists just in this: that he himself is able to set his end in accordance with his own concepts of duty; and it is self-contradictory to require that I do (make it my duty to do) something that only the other himself can do. (DV 386)

In other words, it is practically impossible to promote the perfection of another person. Virtue is a function of moral awareness and individual choice.[33] You might think that you are promoting his perfection, as a parent might force a child to apologize, but perfection is a function of the free will, and so such a parent is only precluding the possibility of virtue in this case (if the child is old enough to be virtuous). Bowen family systems theory teaches us that we might be continually "overfunction" or "underfunction" in our reasoning capacity vis-à-vis another person.[34] In other words, lopsided relationships can hinder moral development. Promoting the virtue of another only precludes him from doing it himself.

We might think that it is very strange for Kant to make such a distinction between promoting the happiness and perfection of others: if we cannot possibly promote another person's virtue, then why should we worry about overstepping our boundaries? Is Kant himself going too far in writing about moral theory and giving lectures on ethics? As long as people are free to make their own decisions, giving rational arguments to sway them should not be construed as overstepping the boundaries of respect. In fact, we might go too far in the opposite direction and fail to recognize the moral internal struggles of others, hence encouraging them to repress them in conformity with an apparently amoral society. How do we know where the line is between trying too hard to influence another person and not trying hard enough? The question of how far to go in trying to affect people's decision making is one of primary importance for the cultivation of intellectual virtue, treating people as rational beings, and promoting the happiness of others. In striving to navigate the path between the paternalism of trying too hard to sway someone and the patronization of silently judging someone else's reasoning to be flawed, we must use great skill in emotional navigation for ourselves as well as for the other. In general, dishonesty reveals disrespect and honesty, if it is attainable at the time, is a good protector of equality. In many ways, the task of negotiating the need to respect our own reasoning and the need to respect another person's reasoning, while being in communication with each other through weathering conflicts, is the primary struggle involved in forging and maintaining relationships.

Treating people as equals, and overcoming selfishness, is a necessary prerequisite for respect. Kant sees respect as a keeping of one's distance, not literally, but in the sense of remembering that someone else is different and separate from oneself. It is contrasted with love, the feeling that one is united with another, even though respect is also necessary for love.[35] Kant argues that lack of respect takes the forms of arrogance, defamation, and ridicule (DV 465). Even though Kant makes a distinction between moral rationality and pragmatic rationality, assigning moral worth only to the former, he believes that we must respect human rationality in general and the human ability to rationally direct one's personal conduct. The moral requirement to respect the free choice of others has been called "the priority of the right over the good" in Kant's ethics,[36] meaning it is believed that Kant holds that it is more important to protect the rights of individuals than it is to dictate the fulfillment of some notion of goodness, like happiness. Kant scholars have recently come to realize the central role the *Metaphysics of Morals*, with its "Doctrine of Virtue," plays in Kant's ethics and so "the priority of the right over the good" is generally seen as, at least, an overstatement, but the truth in this reading is that Kant cautions respectful distance in benevolence, especially because we want to avoid making the beneficiary of generosity feel indebted.[37]

Still, we cannot overlook the duty to promote the happiness of others. One must not simply refrain from coercing others while seeking one's own goals, but one must work towards creating a moral community, a Kingdom of Ends, whereby we must act as though we are part of a community wherein everyone follows the same principles we enact through our actions. Of course, we cannot count on others doing their part, but if everyone did act together, morally, everyone's happiness would be promoted (G 438).[38] The Kingdom of Ends formulation of the CI is very much like the fully articulated doctrine of virtue:

> The supreme principle of the doctrine of virtue is this: act in accordance with a maxim of ends that it can be a universal law for everyone to have. In accordance with this principle a human being is an end for himself as well as for others, and it is not enough that he is not authorized to use either himself or others merely as a means (since he could then still be indifferent to them); it is in itself his duty to make man as such his end. (DV 395)

The doctrine of virtue is a "doctrine of ends, so that a human being is under obligation to regard himself as well as every other human being, as his end" (DV 410). Here we see that promoting the happiness of others and respecting others are two sides of the same coin: we cannot promote the happiness of others without doing so respectfully, and we cannot respect others without also taking their ends as our own.

Along with the positive ends of promoting our own perfection and the happiness of others, the Kingdom of Ends formulation of the CI, contrary to the way that it is normally interpreted, lays bare the fact that communities are *interrelated* wholes. The idea that we might all pursue our own goals independently and respect others by leaving them alone betrays a psychologically and morally bankrupt notion of humanity. The second and third formulations of the CI establish positive moral ideals that are meant to guide communities and relationships. These ideals are often simplified into negative constraints, but in order to be moral we must also engage in the moral inquiry of discovering how to *actively* respect rational nature, not just in ourselves, but in all people. Respect is not a given. We are not born knowing how to respect ourselves and others. We discover the needs as well as the vices that are universal as well as idiosyncratic as we grow, and we must learn ways to address them.

We can also imagine an idea of a Kingdom of Emotional Ends. The ideal of a Kingdom of Ends is one of a systematic unity of people who are treated both as means and as ends. Imagine a city, let us say, wherein the laws are made by everyone—because they can be seen by anyone to be acceptable. Every citizen, who has his or her own individual aims of life, is also the ruler of the city who can understand the coordinated laws that govern the whole. If we similarly are able to understand the laws of emotion, then this moral ideal is also an emotional ideal. When we are angry at someone, that person is perhaps angry at us too. When we are hurt by someone, it is sometimes the case that that person was previously hurt by something we did or that he or she is in need of sympathy in another respect. This realization does not diminish the importance of our personal emotional needs, but it helps us to see them as a part of a relationship and an interdependent community. Just as we must consider mood and the totality of emotional processes to make sense of a single emotional event, we must consider a community and relationships to make sense of emotional interactions. When we become aware of the ways that our emotions are a part of relationships, we are forced to address other people's emotions in order to fully understand and address our own. The result of this expanded, relational perspective is equality and shared respect. The same thing happens with moral judgments: in achieving a universal perspective we come to see others as equal to us, and we come to see harms as equally bad, no matter to whom they occur. In both cases, this expanded viewpoint makes us more mature. In ceasing to demand special status for our judgments, or for our emotions, we put ourselves in a position of equality with other people. It is this mutual recognition that makes us members of a community. We realize that we are no more valuable, and, what is sometimes more important, yet related psychologically, no less valuable.

A community of rational beings requires intellectual communication, which requires the ability to negotiate disagreement. Respecting people's rational ability to make decisions cannot mean that we treat rationality as though it were fundamentally private.[39] Kant argues that the duty to respect the humanity of every person entails

> a duty to respect a human being even in the logical use of his reason, a duty not to censure his errors by calling them absurdities, poor judgment and so forth, but rather to suppose that his judgment must yet contain some truth and to seek this out, uncovering at the same time

> the deceptive illusion...The same thing applies to the censure of vice, which must never break out into complete contempt and denial of any moral worth to a vicious human being; for on this supposition he could never be improved, and this is not consistent with the idea of a human being, who as such (as a moral being) can never lose entirely his predisposition to the good. (DV 463–464)

This is great advice for teachers, friends, lovers, parents, political pundits...everyone: when you think that someone else is wrong, do not jump to character assassinations; try to figure out where that person is coming from. As easy as this sounds, the psychological reality of engaging our intellectual opponents is intellectually and emotionally challenging, even exhausting. Nevertheless, accepting this challenge builds emotional understanding.

Similarly, Kant advocates forgiveness:

> It is therefore a duty of virtue not only to refrain from repaying another's enmity with hatred out of mere revenge...partly because a human being has enough guilt of his own to be greatly in need of pardon and partly, and indeed especially, because no punishment, no matter from whom it comes, may be inflicted out of hatred. –It is therefore a duty of human beings to be *forgiving*...But this must not be confused with meek toleration of wrongs...[or] renunciation of rigorous means...for preventing the recurrence of wrongs by others; for then a human being would be throwing away his rights and letting others trample on them, and so would violate his duty to himself. (DV 460–461)

We can see that conflict resolution skills, in their various forms, are important for both morality and psychological health. In both senses, they are difficult and necessary for maintaining respectful relationships.

Kant discusses the course we should take when we are ourselves the unfortunate victim of ridicule, which is never morally permissible. He writes that we should either "put up no defense against the attack or to conduct it with dignity and seriousness" (DV 467). It might seem preposterous not to defend oneself from ridicule, but one must also remember that not everything everyone says is true and we should only concern ourselves with finding the truth. The truth can usually be seen by others easily enough, but when it is necessary to "set the record straight," we must be careful not to return a wrong for a wrong. The topic of responding when we are the victim of ridicule or the way we should conduct ourselves during an angry exchange is not just an emotionally pragmatic affair; it concerns morality and perhaps to the highest degree. Hence emotional wisdom (what might be called emotional intelligence) is an aspect of virtue and necessary for fulfilling our moral duties.

Conflict is a crucible of both virtue and emotional self-improvement, and, in this case, virtue and emotional understanding seem to be the same thing. We can see that emotional egalitarianism dovetails with moral egalitarianism in the sense that both require openness to the emotions of others. Such is the highest form of emotional understanding: who could possibly project emotional health and intelligence more than he who can speak respectfully and intimately with his intellectual opponents without becoming unnecessarily upset or causing offense? Such a person seems like a moral and psychological hero.

We must not forget that respecting oneself and respecting others are intimately related.[40] Furthermore, the antidote to selfishness may be proper self-esteem. It is often not the result of a puffed up sense of self, but of feelings of worthlessness and meaninglessness that cause us to justify selfishness as a fulfillment of the need for value. Thereby, we falsely identify selfishness with caring for the self. Kant argues that basing our self-worth on debasing others will ultimately be unsuccessful. Regarding arrogance, Kant argues that one who expects others to grovel before him would himself be willing to grovel were the situation reversed (DV 466). Indeed, a good part of the development of moral comprehension may involve hammering out the difference between self-respect and selfishness, as we do in chapters 6 and 7.

Attempting to understand one's emotions at all may strike some as selfish because it requires the devotion of time and attention to oneself. Nevertheless, this kind of self-centeredness is required by the duty to perfect oneself. To understand oneself, to be patient and forgiving with oneself, and to stand up for oneself and resist oppression does not constitute selfishness.[41] Again, it is all too often the case that not being able to recognize and vocalize her opinions, preferences, and needs causes a person to feel threatened by the opinions, preferences, and needs of others or to blame other people for the fact that she has failed to respect herself. In such a case, respecting others is clearly not even a possibility, even though this person might seem very giving, even self-abnegating.

Thinking about moral universalism and egalitarianism in terms of emotionality has led us to consider both the development of self-understanding and the aim of morally good intentions, as well as a concern for the emotions of others. It has become clear that having close, emotional relationships and sufficiently caring for the people with whom one is involved is morally necessary, and even the best way to direct the majority of one's efforts in promoting the happiness of others. Moral sensitivity and emotional sensitivity merge at this juncture: one simply cannot promote the happiness of others if one cannot recognize emotional harm, and one cannot recognize emotional harm and health in others if one is closed off to this part of herself.[42] We do sometimes, of course, help others in a more distanced way, by giving money, for example. We might imagine someone who *only* helps others in this way: this person would have no close relationships, since closeness necessarily entails emotional involvement. (Although Kant had many close friends, his own life may have tended in this direction, falling short of virtue.) Kant's argument for the reciprocal need to promote the happiness of others shows us that we similarly need to have and take care of close relationships since we all need them. Since we have some latitude in determining whose happiness we are to promote we might often be tempted to lie to ourselves about the worth of a particular relationship, but doing so is probably nothing but a moral failing.

In some cultures much of the moral work of caring for the psychological and physical health of people falls to women, but such an arrangement is neither psychologically optimal nor sustainable. Instead of having the cultural collapse of close, emotional relationships, all adults must take up the slack of providing the physical and psychological care that we all need. In this way we can further see that emotion is a necessary dimension of universalism.

6

The Path of Vice

Being open and sensitive to moral feelings starts us on the path of virtue, as we have seen, but not all emotions are virtuous.[1] Vice similarly has characteristic emotions as well as characteristically vicious modes of responding to emotions in oneself and others. Nevertheless, it would be misleading and potentially dangerous to suggest that there is anything less than a complex division between virtuous and vicious emotions. There is no third variable, as it were, that gives us a shortcut for a distinction. All we can do is evaluate whether or not certain emotions are vicious or virtuous, as we discussed in the last chapter with emotions that harbor selfishness and lack of respect. One such mistaken third variable is cognitive awareness. One might think that those emotions of which we are most cognitively aware are necessarily the most virtuous and those emotions with latently cognitive or unconscious causes are vicious. That is not the case. In fact, Kant's account of vices—as conscious antimoral obsessions—are at the heart of his psychology of evil.

This chapter gives a broader context for our discussion of moral feelings by providing the other half, as it were, of Kant's theory of emotion. It will also help the reader make sense of the normal interpretation of Kant's theory of emotion, namely that emotions are uniformly problematic. Kant's theory of emotion gets so much bad press that it would be unfair to end any explanation of it after discussing only his theory of moral feelings. When most readers of Kant wish to isolate his treatment of emotion, they look to what he says about affects (*Affecten*) and passion (*Leidenschaft*) (vice) instead of looking to his theory of feelings, which is the more obvious central location for his theory of emotion. In this chapter we will give special attention to potentially problematic (selfish, petty, rash, etc.) feelings and the truly problematic obsessions that are not what we normally call emotions at all but long-term inclinations that shape an immoral character (i.e., vices). (In seeking possibilities for altering these immoral obsessions we must continue to emphasize that psychological solutions rely on psychological knowledge, which is continually being gathered and which is continually being improved.) Readers of Kant often miss that in his discussion of affects and passions, he seems to be suggesting that vice comes from an improper way of dealing with affects, which are themselves benign. We will consider the proper response to affects—one that prevents them from developing into vices. Last, we will meditate on the psychological causes of evil, responding to the criticism that selfishness does not seem to be its sole cause, as Kant appears to assume.

MORAL AFFECTS

Kant does not use any term in German, like "*die Emotion*," that might be easily translated into "emotion." In preparing his English version of Kant's *Anthropology for the Pragmatic Point of View*, Dowdell chose to translate the German word "*der Affect*" (Kant uses an archaic Latin spelling) as "emotion," which is likely the cause of some confusion.[2] In some places Kant writes that *Affecten* are problematic—like an illness—so to translate *Affect* as "emotion" is probably more of a reflection on the translator's negative view of emotions rather than Kant's. It makes more sense to translate *Affect* as affect instead of as emotion since that English word also tends to connote a shallow or problematic feeling.[3] Our English notion of emotion is broader and should be seen to encompass Kant's comments about feelings, affects, and perhaps even some passions (*Leidenschaften*).[4]

Gregor similarly recognizes that Kant's use of *Affect* and *Leidenshaft* are different from the English terms "emotion" and "passion" because he deliberately means *Affect* to refer to "a feeling (e.g., anger) which precedes deliberation and makes this difficult or impossible" and *Leidenshaft* to be closely associated with vice.[5] Kant does identify a few virtuous passions, but these are the exception to the rule. In English, we talk about being in the "throes of passion," but even this is not exactly what Kant means by passion, since Kantian passion is a *long-term, habituated* state, one that never negates culpability. In English it is much more common for us to use the term "passion" to refer to a constructive desire, or set of desires, such as having a passion for music or pursuing one's goal passionately. Follow your passion!—we say; Kant would never say that. To the extent that I can, I will refer to Kantian passions as vices, but should I forget, we must remember that Kant's notion of passion is not the same as the common English notion.

While it might be hard to believe—and therefore hard to see—in the *Critique of Judgment* and the *Metaphysics of Morals*, Kant discusses affects and passions together because he considers it important to distinguish between them. In the case of the *Anthropology from a Pragmatic Viewpoint*, where we find a more sustained discussion of affects and passions, many readers, looking at the divisions labeled in the table of contents (Cognition, Feeling, Desire), mistakenly conclude that affects and passions are both part of the faculty of desire. In fact, affects are *feelings* and are rightly discussed in Kant's section on the faculty of feeling.[6] His discussion of affects actually begins in the first section (on the cognitive faculty) because emotions (including affects) have mental causes. Nevertheless, in the third section of this work ("On the Faculty of Desire") Kant again makes the distinction between affects and passions before he embarks on a more elaborate discussion of passions. This placement causes confusion about the nature of affects—they are actually feelings—and about their relationship to other feelings, especially moral feelings, making them seem instead as though they are of a piece with passions. Indeed, this choice of Kant's to take up a more systematic discussion of affects with the discussion of passions, while he rightly notes that this discussion belongs in the section on feeling,[7] has perhaps played the largest role in leading readers to falsely locate Kant's theory of emotion as a part of his theory of desire and to overlook Kant's theory of feeling, and even his theory of moral feeling, even though he devotes more time to it than he does to affects and passions put together.[8] The confusion might also come from the fact that Kant does not discuss moral feeling much in the *Anthropology*, presumably because he does not want to confuse the reader as to its source in pure, not empirical, reason, when anthropology contains empirical observations.

In our treatment of Kant's discussion of affect and passion we will have to make sense of two things: one, that Kant introduces these terms in order to distinguish between them, and two, that he nevertheless discusses them together. We will also have to clarify their relationship to moral feelings.

First, Kant writes that both affects and passions "exclude the sovereignty of reason." Also, they are "equally vehement in degree," but then he notes that they are "essentially different from each other, both with regard to preventive measures and to the therapy that the spiritual physician [psychologist] must apply" (A 251). Kant writes:

> The inclination which can hardly, or not at all, be controlled by reason is passion. On the other hand, affect is the feeling (*Gefühl*) of pleasure

> or displeasure at a particular moment, which does not give rise to reflection (namely the process of reason whether one should submit to it or reject it). (A §73)

Affect is feeling before it has been consciously reflected on and evaluated; passions are more conscious and deliberate inclinations, but they are, for that reason, even less "rational," that is, moral.

Comparing affects with passion can be confusing because affects are a kind of feeling, a discussion about which we have already been having, and passions are (probably immoral) cognitive orientations. In other words, affects and passions are two entirely different kinds of things. It is not just that affects involve physiological occurrences to a greater extent than passions do; affects are more morally neutral, while passions are almost always morally pernicious.

In considering Kant's philosophy of emotion overall, we might hypothesize that there are spectrums both in terms of how cognitively attuned we are to the emotion and of the extent to which our emotions seem to be virtuous. In making sense of Kant's philosophy of emotion, we will hypothesize that emotions, for Kant, can be made sense of in terms of a Cartesian plane, where one axis represents degree of immediacy or cognitive explicitness and the other represents moral goodness. (See figure 0.1.)

Affects are feelings, and so they are the effect of some other mental event, like a perception, imagining, or rational judgment. (See chapter 4 for a more detailed explanation of feeling.) Like feelings, they involve physiological occurrences. Kant writes that affects make reflection difficult; nevertheless, this does not mean that they are necessarily oriented toward vice. As we shall see, Kant gives examples of virtuous affects. Looking at our Cartesian plane in figure 0.1, affects then occupy quadrants II and III.

We should try to be clear about what we mean by cognitive explicitness: affects often involve thoughts, and it is difficult to determine in any case whether feelings or thoughts are experienced first or whether or not they happen at the same time. Nevertheless, Kant writes that affects make reflection difficult. They might even be prone to twisting or creating thoughts that justify them. Perhaps we can hypothesize that in the case of affects, it is the physical feelings that themselves seem to be leading thought, at least for a brief time, or perhaps it is the thoughts themselves that do not yield easily to reflection. Kant holds that affects strike us and fizzle out quickly, whereas passions last longer and become habitual inclinations toward certain actions. He writes,

> Affect is surprise through sensation, whereby composure of mind (*animus sui compos*) is suspended. Affect therefore is precipitate, that is, it quickly grows to a degree of feeling which makes reflection impossible (it is thoughtless). (A §74)

Kant often assumes that the term "affect" is a negative term. This is because affects are feelings that arise from thoughts that have not been fully worked out and subjected to moral deliberation. If we look at the taxonomy from the second section of the *Anthropology* ("The Feeling of Pleasure and Displeasure"), wherein Kant explains that feelings have either sensuous or intellectual causes, we can see that the feelings that arise from direct experience or the imagination—even feelings that arise from concepts—are still in need of moral evaluation (A 230).[9] Nevertheless, just because a feeling does not arise from rational, moral deliberation (like respect for the moral law) does not mean that it is ultimately incompatible with rational, moral deliberation.

It is hard for Kant to stick to his definition of affects as problematic. We see this in the *Anthropology* where he distinguishes between affects and moral feeling, refusing to call the latter "feeling" but instead a "pathological (sensible) impulse to the good." Nevertheless, he cannot keep up this distinction, and he quickly refers to "an affect that has the good as its object" (A 254). While we must keep in mind that Kant often uses the word "affect" in opposition to moral feelings, referring to something that is necessarily problematic—especially in the discussions that relate them to passions—it makes more sense for us to assume that since affects are feelings, they can be virtuous and related to moral feelings, making instead a distinction between virtuous and vicious affects.[10] Any reader of the *Anthropology* will see that Kant does not list the affects in order to demonstrate that they are all irrational; in fact, the opposite happens, namely he comes to realize that many of the affects are healthful or related to moral feelings. Whereas his list of passions is more consistently a list of vices.

If, in a virtuous person, one's everyday experiences and thoughts are consistent with having a virtuous character, then we can see that moral feelings, which we discussed in chapter 4, might be more or less immediate. In other words, moral feelings must include moral affects. Respect for the moral law follows from explicit rational thoughts. Still, the deliberation that precedes the feeling of respect might be more or less elaborate. If one has gone through a rather long categorical imperative decision procedure to determine the right course of action, the cause of the corollary feeling will be very cognitively explicit, and the feeling that will accompany the enactment of the decision will be more like courage. Awe, on the other hand, or the feeling of the sublime, will accompany the *perception* of morally amazing acts.[11] Enthusiasm, on the other hand, is moral feeling that is rash and potentially confused.[12]

One should indeed have a quick and immediate moral response from the perception of need—if a stranger falls off a dock, for example. These quick responses (affects) could nevertheless lead to genuine moral action, even without time to better evaluate them. They might also lead to moral mistakes—if the stranger were herself saving someone, perhaps.[13] Nevertheless, Kant could not possibly have a theory of virtue without this way to account for *quick* action.[14] In short, it is likely that moral feelings normally occur all along the axis of cognitive explicitness, but more explicit moral evaluation, although perhaps never totally complete, is a continual component of moral deliberation. In other words, any feelings in the top half of the plane (figure 0.1) should be tending to the right or else they risk tending down. Nevertheless, just because a feeling is not the effect of pure reason does not mean that it is vicious.

Affects, even if conducive to morality, are still worrisome for Kant because they tend to make reflection difficult. Although Kant writes that virtue presupposes apathy, he explains that apathy does NOT mean "lack of feeling and so subjective indifference" (DV 408). Instead, for the virtuous person moral feeling (respect for the moral law) is more powerful than immoral affects or passion (DV 408). Kant calls the former "feelings arising from sensible impressions," opposing them to moral feelings, but this distinction is not very helpful because respect for the moral law could be triggered by a sensible impression, from seeing a good deed or a person in need, for example. It would be more helpful to oppose moral feelings to feelings not arising from or mediated by moral principles. Kant writes that it is always "a considered and firm resolution to put the law of virtue into practice" that must lead the way in moral feeling (DV 409). As we saw in chapter 4, even moral feelings cannot be blindly followed but must be continually checked and evaluated. In the case of respect for the moral law, we have both the feelings and the moral, cognitive reasons behind them, so it is not really possible to speak of "blindly following" this feeling. Nevertheless, the moral, cognitive content might be more or less worked out and explicit. In addition, it might lead immediately to action or lead to more moral deliberation. We can see that either effect (action or more thought) is potentially problematic in that it might ignore the other's necessity, and so a virtuous person needs to be prone to both quick moral action and further moral deliberation.

Affective Expression

While I have highlighted moral affects to challenge the assumption that Kant is uniformly critical of affects, there are also vicious affects; although, it is probably not the affect itself that is vicious, but the way we respond to it. One might think that anger is necessarily vicious.[15] It is not—at least, not according to Kant. Perhaps there is such a thing as a necessarily vicious affect, like schadenfreude, for example: the sinister laugh that escapes when another person makes an ass of himself.[16] Natural sympathy would be its virtuous corrective. Nevertheless, while Kant writes that affects make reflection difficult, we must assume that, in most cases, reflection follows nonetheless, and there is not much we can do to change the fact that we have affects. The opposite is in fact the case: Kant most often assumes that we should simply express our affects and that they will pass quickly enough on their own. Learning to sail through the waves of affect is part of a virtuous life.

Of course, this recommendation, to just let affects take their course, is not entirely helpful. A virtuous person might be prone to virtuous, or at least benign, affects, but we would hope that a vicious person would keep *her* affects bottled up, wouldn't we? Well, no. While we should never act without morally affirming our actions, bottling up affects is never called for, merely reflection. Still, some affects are more resistant to moral reflection than others and are thereby more related to passions.

When Kant first begins discussing various affects in the *Anthropology* (in the section on feeling), he notes that oscillating between pleasure and pain causes enjoyment. We busy ourselves with various activities, causing pleasure, to avoid the pain of boredom. Although one might see the cynicism of an existentialist or an ascetic in these remarks, Kant argues for the value of cheerfulness and affective sensitivity. It is clear that Kant sees affects as a normal part of our everyday flow of experience.[17] Nevertheless, we have some control over how we let ourselves be led by our thoughts and, therefore, by our affects, and we must make choices that are healthy and moral (A 230–236).[18]

Kant most often holds that the best response to affects is to do something with them: either let them out immediately (by laughing or crying, e.g.) or transform them into action, if that is what is called for (A 236). Kant gives the example of anger that if not expressed, allowing for feelings to be shared and the problems to be addressed and solved, will become hatred (a passion; A 260). We should not fall into the Stoic trap of thinking that the affect is itself the problem. When we fail to address anger immediately, we become resentful and hateful. In that case, we have failed to respect ourselves.[19] Kant suggests that the best way to respond to emotions like anger and shame is to express them immediately so that they do not turn into resentment. Resentment must then be vented by people "verbalizing their concerns" (A §78). Kant rightly points out that such verbalization is difficult and that the emotions themselves seem to make it difficult. He concludes, "for this reason these emotions present themselves in a disadvantageous light" (A §78). Kant uses the emotion of anger to illustrate, as he believes that anger is quick to strike and quick to pass: "What the emotion of anger does not accomplish quickly will not be accomplished at all. The emotion of anger easily forgets" (A §74). If we are reasonably able to express our anger, it will pass easily enough.

In the *Metaphysics of Morals*, Kant's writes that for virtue, we must *tame* our affects and govern our passions (DV 407—my translation). The notion of taming (*zähmen*) has two meanings here: one is calming, as taming a wild animal makes it calmer and less dangerous, but also training and educating. Mostly, for Kant, taming the affects simply means calming down.[20]

Kant writes that the affects of anger and shame have the peculiarity of making us less capable of realizing their end.[21] Anger thwarts its own purpose; for example, while we are yelling at someone, he is likely to feel scared or angry in return and therefore unable to think about what we are saying (A 260–261). Or, out of embarrassment, we might not explain our outrage to the other. Everyone feels slightly embarrassed at having to say "my feelings are hurt." Perhaps this is a further reason that Kant thinks that affects have the tendency to frustrate the very purposes that they inspire. For example, if someone's actions upset me, the most straightforward response is to tell that person that they are upsetting me and to ask them to stop. To be upset, though, is a negative emotion and difficult to express. The most unemotional person could simply state: "I feel teased; it upsets me; please stop," or make a similar request, but the true-to-life emotional person often becomes upset and stymied by her anger. Hence, it does seem that the negative emotion frustrates its own goal, but, as we discussed in chapter 3, such difficulty is probably not truly the fault of the emotion, but of one's emotional habits and lack of emotional know-how.[22] Nevertheless, we must take note that affect, according to Kant, thwarts its *own* purpose. In objecting that emotions do not effectively serve their own purposes, Kant implies that the purpose, in the case of anger at least—to avert the perceived evil—is constructive

(A §78). If an emotion did not have a worthy purpose, it would be of no consequence that it did not effectively promote it.[23]

Perhaps affects make reflection difficult because they are occasioned by thoughts of which we are not entirely aware in the first place. In the section "On the Ideas We Have without Being Aware of Them" Kant argues that "it is as if just a few places on the vast map of our mind were illuminated." He goes on to write: "This can inspire wonder at our own being, for a higher power would need only cry 'Let there be light' and then without further action... there would be laid open before the eyes half a universe" (A 135). The examples Kant gives of the unconscious activities of the mind are perceptions that we do not fully perceive and beliefs we do not fully affirm. Also, unconscious ideas can be repressed thoughts since "we have an interest in removing objects that are liked or disliked by the imagination" (A 137). Since "[affect] is surprise through sensation whereby the composure of mind is suspended" it is plausible that negative emotions are difficult to reflect on precisely because they involve pain (A 252). It may be the case that we are unaware of the ideas in the first place because we have tried to push them out of consciousness.[24]

Such an account would help explain the reason that emotions do not always match up with their corollary situations; Kant's theory is thereby an improvement on many variants of cognitive functionalism. Aristotle and Kant both discuss the emotion of anger. Aristotle holds that it is a response to the judgment that one has been slighted. Kant agrees, but he qualifies it by explaining that surprise comes from "embarrassment at finding oneself in an unexpected situation" (A 261). I think that Kant is right to suggest that negative emotions are accompanied by some degree of embarrassment and that this furthers the disruption of thought. The causes of the embarrassment are necessarily personal, but without this suggestion that we have become aware of something that we would prefer to hide, such as our own insecurity, we cannot explain why some slights are angering while others leave us unscathed.

The conclusion then seems to be that the best response to affects is to express them in the most accurate way possible. Still, one is tempted to conclude that one should still try to interrupt the affect's tendency to cause behaviors without any sort of mediation from reason. Sometimes the behaviors that affects cause are clearly immoral: A bus passenger hearing two women making fun of her clothes behind her back might turn around and spit at them. If someone cuts in front of me in line at a movie theatre, I might curse at them. If a student sees a classmate bullying another person, he might punch him. These reactions are all problematic, but despite appearances, it is not their immediacy that makes them wrong but their content.

Kant writes that quickly discharging anger is most likely the best course of action, and that is true, if one does so in a way that is respectful and effective. In the case of the two women mocking the clothing, the fellow passenger most likely should react immediately, only in a different way. She should say something that is both true and to the point. If she is upset because she thinks her clothes are stylish, she should say so. If she is upset to be the victim of belittling, she should say so: "No matter how ugly my clothes are, I am still a human being and upset that you have mocked me." To be afraid of doing so is cowardly, disrespecting oneself. To pretend that the other two do not deserve to be spoken to is spiteful and disrespects them. If someone has cut in front of me in line, I should let that person know that or, if he is insensitive to my appeal for justice, I should let the manager know. If someone is being bullied, the onlooker should come to the aid of the victim, but in a way that does not increase the violence. He should call on moral rules: "Stop! Everyone deserves respect, and this person is being hurt." Even if one cannot come up with the right words, the ensuing back-and-forth will help clarify the emotion and express it.

Baxley argues that Kant theorizes three different approaches to affect: containment, maintenance, and cultivation.[25] For summarizing Kant's treatment of affect (not passion), containment is not the most accurate term. As we saw, Kant recommends "discharging" affects so that they do not build up and turn into obsessions. If we contain them, they might just go away on their own, but Kant assumes that they will not. Still, as a general strategy, expressing all of our affects is probably not wise. An affect could involve a strong inclination to do something immoral, as passions are. Containment is most likely necessary sometimes, but in these cases the cultivation of different responses in the future is more important since containment never solves the problem.

Many of us, when we are hurt, angry, or scared are most likely to lash out in violence. So how do we become the sort of people who are prone to virtuous responses? We must contemplate our moral duties when we are not in an emotional state so that following through with them will be second nature when we are. If we steadfastly hold that all people deserve respect and care, regardless of their past behaviors, for example, this belief will not be thrown out the window when someone attacks us in anger. Pondering our duties should even make us look for opportunities in which to discharge them. Perhaps keeping in mind that we ought to express our affects will actually help us to formulate the most effective forms of expression; whereas telling ourselves that we will hold them in might actually lead to their expression in ways that have been less fully considered.

There are certainly a number of examples of cases in which people appear to act irrationally *because* they are in the throes of an affect. Making use of its purely derogatory meanings, Kant agrees with the Stoics that "the prudent man must at no time be in a state of affect." Nevertheless, Kant mostly thinks that affects go away by themselves. Kant gives the example of a rich man whose servant breaks a beautiful vase. The man gets angry, but when he "compares the loss of one pleasure with the multitude of all the pleasures that his fortunate position as a rich man offers him" the affect abates. It is almost as if affects are occasions for us to remind ourselves what is truly valuable in life and reorient ourselves morally.[26]

PREVENTING PASSIONS

There is really no need to eliminate affects in one's life, but we should try to stop ourselves from having passions, in Kant's sense of the term. Kant writes:

> *Affects* and *passions* are essentially different from each other. Affects belong to *feeling* insofar as, preceding reflection, it makes this impossible or more difficult. Hence an affect is called *precipitate* or *rash* (*animus praeceps*), and reason says, through the concept of virtue, that one should *get hold of oneself.* Yet this weakness in the use of one's understanding coupled with the strength of one's emotion [*Gemüthsbewegung*] is only a *lack of virtue* [as opposed to a vice], as it were, something childish and weak, which can indeed coexist with the best will.[27] It even has one good thing about it: that this tempest quickly subsides. Accordingly, a propensity to an affect (e.g., *anger*) does not enter into kinship with vice so readily as does a passion. A *passion* is a sensible *desire* that has become a lasting inclination (e.g., *hatred* as opposed to anger). The calm with which one gives oneself up to it permits reflection and allows the mind to form principles upon it and so, if inclination lights upon something contrary to the law, to brood upon it, to get it rooted deeply, and so to take up what is evil (as something premeditated) into its maxim. And the evil is then *properly* evil, that is, true vice. (DV 166)

Affects might be rash, but passions are *evil.*

It is interesting that Kant continually distinguishes between affects and passions; why is this an important distinction to make? They differ in terms of their cognitive explicitness and their duration. One would think that a typical rationalist would argue that "emotions" are less problematic to the extent that they are cognitive. In this case, Kant says the opposite. Maybe the Stoic would suggest that emotions are better the shorter they last—that is a possible interpretation here.[28] Still, the Stoic would argue that affects are still not acceptable because they interrupt the calm of ataraxia. Kant says the opposite: the physical upheaval of many affects is good for health. Kant notes that laughter, weeping, and anger all seem to facilitate a release. (Laughter exercises the diaphragm, aids digestion, and promotes society.) Kant seems to recognize that affects, despite being prereflective, are human and a valuable part of life. In addition, there are morally useful affects, as we have already noted.

Another reading is possible: Kant is intentionally rebutting the Stoic and paving a way for the moral importance of some affects. Kant states that the defining feature of affect is "not the intensity of a certain feeling…[but] the want of reflection in the comparison of this feeling with the sum of all feelings (the pleasure or displeasure) in one's own condition" (A §75).[29] This distinction implies that while affects are particularly intense, so are some moral feelings that are based on better worked-out deliberation. One might think that intensity of feeling is itself the problem, and Kant sometimes suggests this: Kant argues that affects, such as anger and shame, can be "incapacitating because of their intensity" (A §78). Strangely, passions, which are more morally problematic, for Kant, are not intense feelings at all, but cognitive obsessions. It is also possible that Kant is highlighting the psychological attention that must be paid to the road between affect and passion so that affects do not turn into passions. My suggestion is that virtue requires us to pay attention to our affects, but we must pay *the right kind* of attention.

Both affects and passions can influence our behaviors without much conscious moral evaluation, and in this respect they are similarly problematic.[30] Both affects and passions tend to cause us to do things without fully thinking about it. In the case of affects, these actions are undertaken quickly and not subject to any sort of cognitively explicit deliberation (about consequences, e.g.) at all. A good example would be shooting off a quick response to what one perceives to be a snide or dismissive email or text. This sort of action feels good in the moment, but it is not the most likely to bring about a resolution to the original problem. The way that one expresses one's hurt feelings makes all the difference in whether or not the feelings are virtuous. In the case of passion, there is considerable cognitive deliberation—brooding even—but no moral evaluation. Take avarice for example. Planning strategies for making money could fill most of one's time—and one might be brilliant at it—but the goal of making money is never itself evaluated, and it is maintained even at the cost of immoral means.

Nevertheless, while affects are a part of the faculty of feeling, passions are a part of the faculty of desire:

> Affects differ in kind from passions. [Affect] relates merely to feeling, whereas passions belong to our power of desire and are inclinations that make it difficult or impossible for us to determine our power of choice through principles. (CJ 272 f.)

Affects, therefore, may or may not develop into immoral desires; passions are already immoral desires. Moral feelings relate to a special kind of desire—morally caused desires. Still, affects, like moral feelings, might be disconnected from desire.[31]

Kant's criticism of passion is the same as his criticism of affect, but Kant takes passions to be more recalcitrant to moral reflection: "inclination, which hinders the use of reason to compare, at a particular moment of choice, a specific inclination against the sum of all inclinations, is passion" (A §80). Kant uses similar terms to describe the problem of affects, but affects are short-lived. Passions have more power to preoccupy us and continually lead to immoral decisions. Frijda lodges this criticism at "emotions" with his "law of closure":

> Emotions tend to be closed to considerations that their aims may be of relative and passing importance. They are closed to the requirements of interests other than those of their own aims. They claim top priority and are absolute with regard to appraisals of urgency and necessity of action, and to control over action.[32]

Frijda writes that emotions similarly shirk a consideration of consequences. Kant, on the other hand, is less hard on affect than he is on passions. Passions seem to remain in this myopic state, while affects pass out of it. Also, Kant would not lodge this criticism against moral feelings in the first quadrant (figure 0.1) at all.

Kant defines passions as being cognitively explicit and longer lived; he usually assumes that they are immoral (quadrant IV). When Kant writes that passions cannot be controlled by "reason," he does not mean that they are unthinking—they can be ingenuously calculating.

He means that they oppose moral deliberation. Instead of feelings, passions are inclinations or habituated behaviors. Perhaps "obsession" is a better translation than "passion." Kant is not (yet) referring to anything that might be considered a noble passion, as with a passion for truth, justice, or a passion for the arts. Kant's list of the passions (ambition, lust for power, and avarice) is a list of vices (A 268–270). Passions, for Kant, are those preoccupations that cause the privileging of one inclination over all other concerns, and, for this reason, they are essentially immoral.

A passion, for Kant, turns out to be something like an addiction because it imposes the priority of a myopic drive over all or most of our other interests, both pragmatic and moral. It is a false value, a habit—intellectual and behavioral—that structures your life. The moral meta-maxim, which Kant stipulates in his *Religion* essay, is to always follow the moral law over inclination whenever the two conflict; the immoral meta-maxim is to follow inclination (even occasionally) regardless of whether or not it conflicts with the moral law (R 24). Passion, being preoccupied with itself, forces the subject to choose on its behalf over the moral law, should the two conflict, and it therefore sets the subject up for immoral behavior. Kant often uses the term "inclination" (*Neigung*) in this way, as something necessarily opposed to morality, but inclinations are merely partial and perhaps shallow manifestations of moral reason; passions actively oppose morality.[33] A passion, for Kant, is something like a drug addiction; we can very effectively strategize ways to get the next fix, but the fact that this goal takes precedence over all, or many, of our other goals, is irrational.

A caveat: Kant includes the natural passion for freedom in his discussion, describing its natural, moral cultivation, and he does not criticize it in any way. Kant believes that moral thinking helps the natural inclination for external freedom develop into the concept of justice. He describes this transition as a strengthening, not a sublimating, of the passion. In the case of freedom, reason and passion work together: "reason alone establishes the concept of freedom and passion collides with it" (A §82).[34] So, we can conclude that it is possible for *natural* passions to move up from the fourth quadrant into the first with the help of moral reason—perhaps by becoming more universal, that is, including a consideration of others into one's thinking. Kant writes that the desire for vengeance first originates from a reasonable source: the perception of an injustice (A 270). Vengeance is the natural passion for freedom and justice gone awry; it is merely selfish. Here is an example of a passion that can be educated with the sort of moral evaluation that we discussed in the last chapter. If the perception of injustice is accompanied by a notion of universal respect and fairness, it becomes the passion for justice. If it is accompanied by selfishness, it becomes the passion for vengeance.

Natural passions include the desire for freedom and the desire for sex. Both of these can be sublimated into culturally, even morally, beneficial forms. Among the acquired passions, Kant includes the mania for honor, the mania for power, and the mania for possession. "They are inclinations that have to do merely with our possession of the means for satisfying all the inclinations that are concerned directly with ends" (A 270)—"possession of the means to whatever purposes we may choose." In other words, they are all variants of a selfish, maniacal fantasy of our power and worth.

These vices come from false moral ideas. They are all passions for power that come from privileging prudential interests over moral interests. The passion for honor comes from falsely identifying the source of value as other people's opinions instead of morality. The same mis-valuing comes with greed. The origin of the passion for domination is the fear of being dominated by others. Kant counts the obsessions that can develop from any kind of hobby as vices. This is a useful characterization. Kant calls it "the illusion to mistake the subjective for the objective" because in these cases one thinks that some activity that brings some degree of pleasure has inherent value and itself gives life meaning (A 275).[35] The pleasures that accompany collections, games, or other diversionary activities can indeed be useful for physical, psychological, or social goods, but they do not in themselves have value or give meaning to life, although we easily forget that. Passions are obsessions with false goods. "Passion always presupposes a maxim, on the part of the subject, of acting in accordance with an end prescribed to him by his inclination" (A 266)—"making a *part* of his end the *whole*" (A 266).

By the time a passion has developed, it is almost too late to free ourselves from them: "[T]hey are, for the most part incurable diseases because the patient does not want to be cured and shuns the rule of principles, which is the only thing that could cure him" (A 266). A passion is like chains that cannot be broken loose from because "they have already grown together with his limbs, so to speak" (A 267). A more radical life change is necessary. Hence changing one's passions is far more difficult than calming one's affects.

Our Cartesian plane helps us here because it suggests that emotions can go wrong either in that they are too rash or because they are immoral. In some cases, we might need to cool down, but in others might need a totally different approach. We have already discussed various means of cultivating and strengthening moral feelings in chapter 4; it is probably the case that attention to moral feeling is also the best way to prevent affects from turning into immoral passions.

It is also likely that, as with the case of anger and hate, passions can form from incompletely expressing our affects and not seeking their ends in the most effective way possible. For example, with the passion to control and dominate others, Kant writes that it is probably the effect of fearing domination oneself. A more rational response to the fear of domination would perhaps be to evaluate its legitimacy and to avoid aggressive and dominating people . . . not to become one. One should defend and stick up for oneself, but not to the extent of disrespecting the agency of others. Respecting the agency of others seems like a safer strategy for achieving self-protection. Finding ways to express and follow through on our feelings, while morally evaluating them, is the best way to prevent their being developed in impractical and immoral ways. Being disconnected from (or repressing) our own moral emotions, including those feelings that give life to our most important relationships, casts us adrift to find, or even invent, new values—not to mention charting the self-caused storm of affects that is involved in the original obfuscation.

WHENCE EVIL?

Evil has become a popular topic among academic philosophers. Robert Louden speculates that this trend perhaps followed the terrorist attacks of September 11, 2001, and the reference to evil in the popular media that followed thereafter.[36] The dawning public awareness of global climate change also most likely has most people looking to their consciences. Kant's theory of evil has meanwhile gained some attention and scrutiny. Commentators have noted that Kant's *definition* of evil seems to be the most accurate. He defines evil in human terms—even coining the term radical evil—as the choice of some other end over the moral law.[37] Nevertheless, over and above offering a definition, Kant's theory of evil has met criticism because he appears to explain the genesis of evil purely in terms of selfishness.[38] Since my explanation of our moral duties has similarly tended to focus on overcoming selfishness, it is befitting that I address this potential criticism.[39]

Critics are right both that self-love plays the shining role as cause of evil in Kant's philosophy and that many evil acts, such as the airplane hijackings of 9/11, do not appear to be selfish. The fact that the jihadists were willing to martyr themselves for their cause and that they believed themselves to be morally righteous severely complicates any explanation of evil that focuses on self-promotion. Yet, how can someone who is murdering thousands of people actually think he is doing something good?

Louden is right to point out that "self-love" in Kant's philosophy is a more general notion that encompasses any inclination that opposes the moral law, but this line of defense merely retreats into a tautology and does not thereby strengthen Kant's account. In other words, if self-love simply means "whatever motivates immorality" then Kant has only told us that immorality is motivated by whatever motives immorality. Nevertheless, Kant scholars have shown that Kant has much more to say about the moral psychology of evil than the recent spate of critics admit.[40]

If we wish to accept his theory of morality, we should also make sense of evil in similar terms. Following his lead of making evil the result of an individual's choice, we should, as Louden does, look to Kant's account of psychology (anthropology) and to ourselves. The foregoing discussion of passion and vice has shown us many paths to evil along the lines of false moral ideas.

While Kant does explain all evil in terms of choosing against morality, he also introduces the idea that there are three different kinds of evil, in ascending severity (R 29–30). The first level of evil is frailty. The frail person knows what he or she should do but simply lacks some kind of necessary strength, like courage or social skills. The second level of evil is impurity. The impure person does not represent her duty to herself completely clearly. Perhaps she is confused, or perhaps she has succeeded in rationalizing her immoral choice.[41] The third level of evil is depravity. The depraved person incorrectly represents the moral law to himself. Despite these three levels of evil, we should not forgive ourselves evil acts, but see as Kant does, one instance of choosing against the moral law as always radically evil. Remembering this likely evil in ourselves will prevent us from becoming self-righteous.

Frailty and impurity are easier to make sense of in terms of self-love. For myself I can say that oftentimes when I knowingly choose against the moral law it is for the sake of convenience—the all-American false god. There are perhaps a number of justifications that go along with this choice, like the idea that my choices will not have much of an *effect* anyway. In other words, I translate moral considerations into pragmatic considerations. This elision is exactly what Kant means by self-interest.

Gressis is correct to point out that Kant highlights the mistaken valuing of oneself in terms of social values instead of moral values.

We can see this with the vices of envy, ingratitude, and malice (DV 459–460). Drawing from Kant's lectures, Gressis outlines two "moral fantasies" or false moral ideas that explain evil (and both in terms of self-love). The first is what he calls "The Adequacy Fantasy," which is the impulse not to judge oneself too harshly. This falsehood seems completely widespread and perhaps, given the psychological law of the reduction of cognitive dissonance, even somewhat natural—although still not good. This person is also prone to comparing herself with her less than virtuous surroundings instead of to moral ideals. One might even think that this excuse boils down to "peer pressure." (We will discuss proper self-esteem in greater detail in chapter 7.) The "Exceptionalist Fantasy" is the false belief that one is morally great. Gressis again ties this with comparing oneself to others instead of to the moral law. This explanation that he highlights perhaps is best linked to the first two levels of evil. The person who fancies himself to be exceptional might do so because he "frames for himself the idea of an indulgent moral law" (L 348). Gressis also mentions Kant's notion of timorousness, or opting out of morality all together. Perhaps this idea can help us make sense of the third level of evil.

Have we begun to make sense of the evil of the jihadists who wish to kill for the sake of virtue? Kant would be right, I think, to call this "hate," which he often discusses as a passion. This case helps us to see additionally that hate is usually accompanied by self-righteousness. Hate, even nonmurderous varieties, is evil because it debases and disrespects a human being; it is a psychological form of murder. While it is wrong to turn to violence out of hatred, I would argue that the victim of hate also has some moral obligations in the situation, including not to hate in return. If one suspects that she is hated, she should ask whether or not she has wronged the other person and consider respectful ways for addressing his ire. Behaviors of people in a relationship are linked; international terrorism teaches us that we are also responsible for our international relationships.

There are different reasons for hating someone, but usually they all involve some form of danger to self, real or perceived, caused by the other. Hate is the result of anger that has not been properly addressed. We can learn about the evilness of hate by considering its antidote: the proper solution to anger. Anger must be addressed in a way that respects all of the parties involved, and its response can thereby go astray either by not respecting the subject or the object of anger. Perhaps Kant is right to say that sometimes it is best to express it quickly. At least then the person at whom one is angry becomes aware of the anger and is given the chance to redress it. Nevertheless, usually the expression of anger is met with more anger, and so some rational calculation is required in order to truly serve the purposes of the angry person. Merely forgetting the anger is not advisable as it does not respect the dignity of the person who was slighted. Perhaps fittingly, burying one's anger, thereby disrespecting oneself, is also the most likely cause of disrespecting the other since the buried anger usually becomes a "writing off" of the other person turning her from a human deserving honesty and kindness into a mere "acquaintance" thought to deserve little more than avoidance and the semblance of "respect." Nevertheless, as we saw in the previous chapters, the moral notion of respect does not merely involve leaving others alone but giving others as much value as we give ourselves. Here we can see a similar tendency to justify an evil action in terms of a falsely pragmatic understanding of morality—one that reduces morality to legality.

We must notice that heinously evil acts involve failing to see humans as humans, as Arendt notes that evil seems to involve "making humanity superfluous."[42] Stifling moral feeling within oneself is similarly a form of denying humanity. The three false emotional comportments that we discussed in the first chapter—Stoicism, Romanticism, and Positive Psychology—are each failures of self-respect, which, following the parallel between respecting the self and respecting others that we outlined in the last chapter, can each be connected to failures to respect others. To suppress either one's emotional or one's moral being is a disrespect that then tends to similar forms of disrespect for others. Properly morally respecting oneself and one's emotions—morally evaluating them—puts us on the right path for developing virtue while the contrary helps to explain the development of vice and evil.

Anger for someone who has wronged you in some way is inevitable. To ask humans to overcome hatred in favor of forgiveness and respect—two necessary components of conflict resolution—is exactly asking them to see that the moral law is higher than the self-interest of our very survival. It is perhaps in this existentialist sense of a clutching to our own lives that Kant characterizes all opposition to the moral law as self-love. To accept the primacy of morality is a scary and difficult task that I do not ever expect humans to fully accomplish, and so I expect our world to always be plagued by what Louden refers to as the everydayness of evil.[43]

Making use of Kant's three-tiered explanation of evil, it seems to be the case that frailty, impurity, and depravity are caused either by some kind of weakness in following the moral law or having false moral ideas. There is likely a bit of both in every case. For example, why did individual Nazis kill individual Jews? Often times they were drunk and following orders, perhaps knowing that they should not. Other times they believed that they were justified, that Jews are not human or that it is okay to kill in war, out of revenge, or for the goal of improving society, etc.[44]

While we might be able to stretch the term "self-love" to fit all these cases, it is indeed a bit simplistic for Kant to class all evil as self-love. Perhaps we can equate having false moral ideas with satisfying our desires and adopting what Gressis calls the "Prudential Maxim": the belief that fulfilling our desires will make us happy and that we should promote our own happiness. It might also be the case that someone who is evil in this sense—seeking her own happiness—is actually less evil than someone who chooses evil for more abstract reasons. It seems that in the examples of the most horrendously evil people—something we might call coldhearted evil—we can think of, like Josef Mengele, they seem to be acting on false moral ideas, like a failure to recognize the duty of respect for persons, with a lack of natural moral feeling, like sympathy or affection for children. Josef Mengele seemed to think that medical knowledge was more important than human—Jewish—life and that there was nothing wrong with using people as a means to medical knowledge, even when it caused them obvious suffering. He was called "The Angel of Death," as heinously evil people seem be superhuman because of their complete lack of natural moral feeling.[45] Perhaps this false notion of superiority is part of the incentive to evil. It would be *preferable* to explain such heinous evil in terms of self-love. Actions determined by incentives of self-love seem to be explainable as mere weakness of will, what Kant calls frailty, rather than having a radically evil will. Without such an explanation, we find it difficult to make sense of such a person's actions. Without natural moral feeling or a psychological explanation of his having lost such feelings, such a person would seem entirely immune to moral reasoning and would be beyond hope. Even to consider a human that way is immoral since it posits him as inhuman—not human. In fact, it seems that only this variant of evil deserves to be called "radical evil"; for how could one knowingly choose against the moral law without deceiving oneself in some way?

Kant often argues that all human beings are sensitive to the moral law. He calls it a fact of reason. While I cannot speak to the empirical truth of this, if we deny this assumption, at least two untoward conclusions follow. One, the evil person is not ethically responsible for his evil. If he has no idea of the moral law, he is not bound to it, and while he might be punished by positive, governmental laws, we have no moral argument to make against him. Two, if a person has no moral consciousness and no moral feeling, he would not deserve respect. As we saw from chapter 4, the respect that is due to persons is a function of their moral existence, regardless of their behaviors. If we postulate that a person is a moral monster, all manner of treatment of him would be justified.[46] If we are wrong, the effects would be horrendous. Therefore, for the sake of our own culpability, it is safest to assume that people do have a moral core or at least a moral potential. To write someone off, again, seems like itself an evil. While the victims of evil are no doubt tempted to retaliate in vengeance, their victimhood does not excuse immorality.

Aristotle's discussion of the common, but false, ideas about happiness might help us even further. As we saw, Kant covers similar ground in the beginning of the *Groundwork*. Fame, money, pleasure, all of these things are taken by some to be the purpose of life. Perhaps the pantheon of false happiness is as broad as there are possible goals, and so someone like Mengele is merely akin to someone who myopically pursues health or beauty, only he falsely believed medical knowledge to be the purpose of life. In this case, it seems relatively easy to show that these things have only relative not inherent value. If the above account of the generation of passions is at all psychologically accurate, we have begun to show how such false moral notions can be prevented. It shows us how we can and must avoid becoming dead to moral feelings. I do not know if heinous evil can fully be explained in this way, but the majority of everyday evil can, and so we turn to, in the next chapter, a discussion of the purpose of life, moral commitment, and genuine, moral self-esteem.

While we all start out selfish to some degree, we all also are naturally sensitive to moral feelings. The direction in which we develop is left undetermined. One must assume that one's social conditions, most especially one's upbringing, affect one's ability to have moral ideas and moral feelings. Can an account of moral development help us make sense of the heinously evil person? It does seem to be the case that heinously evil people are often the product of bad childhoods. Even still, stopping a moral evil or working for good, always involves going against one's society to some degree.[47] Looking at war, slavery, rape, and child abuse, throughout history, we can conclude that false moral ideas are more common than not. Nevertheless, throughout history there have been some people (however rare) who have displayed moral feeling and moral ideals. The question, to which we now turn, is how to be more like them.

7

The Inner Life of Virtue

Moral Commitment, Perfectionism, Self-Scrutiny, Self-Respect, and Self-Esteem

Having just examined the way that vices are formed, it is incumbent upon us to examine the contrary process since virtue is not merely the absence of vice, but the positive striving for good. We will here consider Kant's view of what a good character looks like: it is what I have heretofore been calling "living a morally committed life." If there is an argument of this chapter, it is a simple tautology: it is good to be good. Nevertheless, because, as we shall see, it is hard to be good, there will most likely always be people who resist this simple conclusion. In this chapter we will discuss what it means to be good, that is, to live a life committed to moral goodness—we will explore the psychological dimension, as well as the benefits and difficulties, of such a life—and we will defend the virtue of this life against possible detractors. The goal is to provide a picture for the reader of what it feels like on the inside to live a morally committed life. We will see that, while it does not make sense to try to live a good life in order to get some kind of selfish reward, it does turn out that moral commitment, while simultaneously involving the possibility of self-sacrifice, does lead to the highest form of self-esteem.

Let us begin with an illustration of virtue since perhaps it is commonly assumed that normal, everyday people are already virtuous enough. The Greek term for virtue, *arête*, means excellence, and so, almost by definition, it is difficult to be virtuous, and average people are not excellent. Kant describes virtue as fortitude, or a sort of courage or struggle, emphasizing the difficulty involved (DV 380). Our moral duties to ourselves require holiness (excellence) of motivation and perfection of fulfillment; of course, since we are not perfect, the best we can achieve is merely striving for perfection (DV 446).[1]

Moral excellence is not easy even for the greatest among us. One of my favorite moral exemplars is Diane Nash from the civil rights movement in Nashville. During a rally, Nash mounted the steps of Nashville's City Hall to ask the mayor face to face: "Do you feel it is wrong to discriminate against a person based solely on the basis of their race or color?" The mayor admitted that he did.[2] Even still, she reported that before going to the lunch counter sit-ins she felt scared to death.[3] Given the beatings and lynching that preceded and followed the actions she helped to organize, she was right to feel that way. Still, she showed great courage, and, with others, she brought national attention and progress to the segregated south.

Indirectly, through James Lawson, Nash was a student of Gandhi. Gandhi was a leader of India's independence movement, fighting not only injustice inflicted by the British but also the Indian caste system and ethnic hatreds. He promoted a life that respected all beings by being committed to following the truth of goodness and nonviolence. He often strategically risked his life to promote the goal of a humane society, and he was eventually assassinated by a Hindu nationalist extremist.

My point is not that Diane Nash or Mohandas Gandhi was perfect, but they were good. It is not easy to be good; it is difficult and rare. It was never easy for these two moral exemplars and it will never be easy for us who have comparably far less virtue and skill; nevertheless, this notion of striving for perfection is inherent in the idea of goodness.

MORAL COMMITMENT AND PERFECTIONISM

Kant is a perfectionist: "For really to be too virtuous—that is, to be too attached to one's duty—would be almost equivalent to making a circle too round or a straight line too straight" (DV 185). Nevertheless, we will have to be careful about the way in which we interpret this perfectionism.

Kant's notion of living a morally committed life is expressed in his central notion of having a good will. Goodness, for Kant, can be explained in terms of having a good will: "Nothing can possibly be conceived in the world, or even out of it, which can be called good without qualification except a good will" (G 393). This is the first sentence of Kant's first published work devoted explicitly to moral theory, and the term "good will" is even in bold face type. The only problem for students of Kant is: what does "good will" mean? Novice readers of Kant will first think of charity or beneficence when they hear the term "good will"—that is, after all, the way we most commonly use the term in English.

Kant scholars have done a good job interpreting Kant's notion of *wille*.[4] It refers not to an individual act of willing, but the entire faculty of willing; indeed, in some ways Kant uses the term *wille* synonymously with the entire faculty of practical reason. There are two senses of the faculty of practical reason for Kant: empirical and pure practical reason (the first is determined by things we experience; the second is determined by pure reason). The term *wille* seems to unite them. Empirically, we phenomenologically consider what it is we are going to do; we weigh our options, consider strategies and consequences, and we make choices. We are affected by physical desires and needs and these in turn affect our decision-making. Nevertheless, we also comprehend the simple truth of morality; we feel its weight, and we feel that it trumps our other desires. Kant's notion of *wille* encompasses all of our decision-making, including both moral and amoral concerns. Therefore, to have a "good will" does not mean the same thing as merely making the right choice; nor does it mean the same thing as having a good intention. It means something more like having a good character: having one's entire practical capacity for making choices be oriented toward goodness.

Kant's discussion of having a good character in the *Religion* essay helps to explain what it means to have a good will. There he explains that there are two possible practical orientations in life: either one places moral concerns above pragmatic, amoral concerns all of the time or one places pragmatic, amoral concerns above moral concerns some of the time. For Kant, it is not permissible to even occasionally compromise morality. Even to do so one time is wrong, by definition, and the willingness to compromise on moral principle is woven into our character. To waver in one's commitment to the good is to not be committed to the good. To have a good will then means to orient one's thinking in terms of prioritizing moral considerations.[5]

In the "Doctrine of Virtue," Kant explains that we have the duty to perfect ourselves both naturally and morally. Natural perfection involves the perfection of all of our natural capacities, like physical health, psychological health, intellectual, as well as artistic and social, excellence, and so on. Moral perfection has two parts: we must do the right thing (that which is morally required of us; both our positive and negative duties), and we must do these things for the right reasons. The second part involves comprehending our moral duties and so this facet of morality, the duty to perfect oneself morally, is the necessary foundation for all of the rest of morality, for if we did not comprehend moral duty we could not be moral, even if our actions happened to be morally acceptable.[6] Moral commitment therefore requires self-knowledge. Even though complete self-knowledge is impossible, Kant writes that "the first command of all duties to oneself" is: "know (scrutinize, fathom) yourself" (DV 441). We will further discuss the necessary extent of this self-reflection shortly, but first we must finish up with our characterization of Kant's perfectionism.

Although duties of virtue are wide and imperfect duties, Kant writes that the duty to increase our moral perfection is "narrow and perfect in terms of its quality" (DV 446). He goes on to say that it is only because of "human frailty" that this duty is "wide and imperfect in terms of its degree." We might paraphrase this as follows: we have the duty to be perfect; we must do what is right and for the right reasons always. Nevertheless, while this goal is theoretically possible, it is nearly practically impossible, and so we must pursue it as best we can. The way that we pursue it is up to our best judgment.

We should be careful about what this means since there are a number of ways one might misunderstand the requirements of morality. We have seen already that morality requires valuing all of humanity; it does not require serving others at the expense of respecting yourself. We have also seen that morality means being fair and not valuing your own pursuit at the expense of others. It does not mean completely sacrificing all personal pleasure. Indeed, it would be a moral failure to forsake one's own duty to naturally perfect oneself. We have seen already in previous chapters that our moral ends are the happiness of others and the perfection of ourselves. There are many ways of promoting these ends, even many necessary ways that might not at first appear to be connected to virtue, like leaving other people alone and taking a nap.

Nevertheless, having a moral orientation in life describes a steady state of one's character not something that we do some of the time. Therefore, it would be a mistake to think that we sometimes do moral things and other times do amoral things. Since morality's demands are always in effect and one should always comply with them, a virtuous person is always thinking in terms of moral acceptability and moral goodness. We might not always be going out of our way to help others or perfect ourselves, but we should always be open and sensitive to the possibility that now might be a good time to do something that helps others or helps us to learn as well as on the lookout for behaviors or inclinations that might be morally prohibited.

Louden describes perfectionism, a term he admits might turn some people off, as the position that, with morality, more is always better.[7] Kant writes that the moral command with respect to one's moral ends is "Be perfect" (DV 446).[8] This approach to morality might strike some as overly taxing, but if we see morality as an orientation, a character trait, that one either maintains or does not, and not as a tallying up of a number of discrete moral deeds, then the injunction that we must cultivate a good will makes sense. Of course, with perfectionism comes infinite striving since, as imperfect beings, beings who can never entirely know our own character and, furthermore, beings who are always faced with new moral difficulties, we cannot achieve perfection. Following Aristotle, not the Stoics, Kant takes virtue to be a stochastic skill; in other words, virtue is a way of orienting oneself and setting goals, not necessarily a way of reaching those goals.[9] Virtue characterizes a way of aiming, not a way of achieving. (Kant's conviction that we cannot achieve perfection is a premise in his—admittedly strange—argument for the necessity of our postulation of an afterlife.)

Even understanding this, many moral theorists still allege that Kant places the bar for morality too high, arguing that a moral theory should have a category of the supererogatory (those actions that are good but not required by duty, and, hence, optional).[10] There is a genuine disagreement between Kant and the supererogationists, who believe that moral duty is something with which we should sometimes be finished and that morality is a constraint on our lives that is at the same level as other constraints, such as needing to work or complete other chores, and must be negotiated accordingly.

Although it is the legacy of liberal Kantianism, which privileges negative, legal duties, that inspires the supererogationist view of morality, as we have seen, Kant understands morality and moral duty in terms of virtue and character—the necessity of developing a good will—which involves the orientation and development of all of our human abilities and behaviors. Character is something that underlies all of our choices; it is not a task among others, but the way that we approach all tasks, indeed, our lives in general. To ask that we might sometimes be able to leave off with moral duties is simply to misunderstand the nature of virtue. Moral commitment is the condition for the worth of happiness, and our lives in general; it is always in effect. Also, given the fact that human consciousness is deep and complex, I am not willing to give up on the possibility that moral commitment characterizes a way of thinking that is always active. (Soon we will explore Kant's notion of conscience.)

One might think that the meaning of "wide" and "imperfect" duties is that they are more lax.[11] The opposite is closer to the truth. Duties of virtue are wide and imperfect because they refer to (overarching) ends, not to specific actions. They are wide because discretion is required for fulfilling them. They are "imperfect" because their content is not specified a priori, unlike with the perfect negative duties, not to lie, murder, etc. Similarly, they are imperfect because any certain action may or may not be morally necessary in a given circumstance. Discretion is required for judging which duties are in play at a particular time (to perfect oneself or promote the happiness of others, for example). The duties to oneself and others correspond to principles that are always a part of a virtuous character. Applying to our inner lives, virtue necessarily involves practical reason—the *wille*; in other words, being virtuous is a product of thinking for oneself. Of course, the importance of thinking for oneself does not mean that anything goes, but that every situation will be different, and the practice of judgment day by day is a necessary part of living and having relationships.[12]

Kant does not offer complete guidance for deciding how much of our lives we should devote to others. In the *Religion* essay he states that we should do as much good as we can.[13] Still, the fact that moral duty is always in effect does not mean that we must devote all of our time to cultivating self-perfection and promoting the happiness of others. Instead, it means that we must always be *the kind of people* who cultivate self-perfection and promote the happiness of others. I cannot possibly even begin to give a more specific answer to the question of "how much?" We should think in terms of quality instead of quantity. Even when we are not doing something that is easily recognizable as perfecting ourselves or helping others, the maxims to do so are still active in our lives, informing all other actions and values. Indeed, to have a moral orientation in life means to put moral goodness first, and when we put moral goodness first, all other possible goals and desires become necessarily less important, as they should. Instead of morality needing to justify its place in our lives, all other pursuits must prove themselves morally worthy.

The very idea of having an answer to the question of "how much do I have to do in order to fulfill my moral duties?" as the supererogationist would like, should appear strange when we consider our normal, moral engagements. I cannot imagine a life wherein there is such an answer. It seems as though having an answer would entail that one stops *feeling* moral demands. Not feeling moral demands would undermine moral feeling and the respect for the moral law—indeed morality itself—altogether. The supererogationist desire to stop feeling moral concern after a certain point strikes me as championing insensitivity and ignorance. It seems that, instead, we will always feel that there is more to do, and that we will always feel regret for not being able to do more.[14] This feeling goes along with the earnest attempt to do all that we can and causes us to continue to ask ourselves whether or not we could possibly do more. We feel the needs of other people and ourselves, we are morally commanded to feel these needs, and, yet, we cannot address them completely. This is a difficult place to be, and we must accept the difficulty.

Understandably, this is a hard reality for people to live with. Perfectionism yields the conclusion that we are imperfect. It also yields the conclusion that there is more that we can and should be doing. We will have to give up a lot of superficial pleasures and conveniences either because, on closer inspection, they are immoral or simply to make time and resources for more meaningful pursuits. Moral commitment also puts us in conflict with those dimensions of culture and reality that are immoral. It would be much easier, and more peaceful, if we could just smile and go with the flow once in a while.

Wolf characterizes moral commitment derogatorily, referring to the "moral saint" as the person "whose every action is as morally good as possible."[15] Following Adams, Louden tries to defend moral commitment by arguing that there are a plurality of different virtues and that it does not make sense for us to tally up all of the goodness of a person.[16] While I agree with Louden that there are a plurality of moral values, it seems that Wolf is right to think that moral commitment involves trying to make one's every action as good as possible. To clarify what is at stake we might ask the question, for example: Would Gandhi ride a roller coaster? I am assuming that riding a roller coaster does nothing but offer fleeting, superficial pleasure to no one but oneself (or those people on the roller coaster). There are, to my lights, three different answers to this question: (1) No; (2) Only if he perceived it to be necessary to fulfill some good, as relaxation is necessary for reinvigoration; and (3) Yes. To the reader attempting to imagine Gandhi riding a roller coaster (perhaps as a humorous greeting card), the question is posed: Why would Gandhi be riding a roller coaster? When you live your life tuned into the reality of suffering in the world and when you have committed your life to making the world a better place, riding a roller coaster *is not even pleasurable*. Why would Gandhi turn his back on the starving peasants and slum dwellers? Why would he turn a blind eye to resolving Hindu-Muslim violence? If there were anything he could possibly do to help, he would want to do it right away, not after he had wasted ten minutes riding a roller coaster.

Of course, we have not ruled out (2), and moral commitment involves a lot of diversion that is instrumentally related to pursuing moral goodness. In many cases, we simply do not know what we can do to help, and sometimes going about our everyday lives will clear our thinking enough for us to get an idea. Nevertheless, taking a break from wracking one's brain about the best strategy for making the world a better place does not constitute taking a break from one's duty to make the world a better place.

Wolf thinks that not being willing to, or not wanting to, take time away from making the world a better place in order to do something that is morally meaningless is a personality flaw. It might just be the case that, as we see at the end of this chapter, this line of objection is merely an immoral means of deceiving oneself about one's moral duties and the sacrifices that are necessary to fulfill them. Or perhaps Wolf is confused about the variety of our duties, which includes, among other things: family and parental duties, duties to oneself (self-respect), and duties to friends.

Perhaps we can make a constructive analogy with parenting: while a parent often takes a break from parenting by seeking childcare, she must never take a break from the duties of parenthood, namely making sure that her children are well cared for. Similarly, while what we are doing might not always look like perfecting ourselves or promoting the happiness of others, we are always on the lookout for ways to be virtuous, as well as skeptical about whether or not we are really trying hard enough. A similar analogy can be made with the constant nature of marital commitment, not just to avoid adultery but to make sure that one is constantly being a good spouse and harmonizing one's thoughts and desires with marriage.

We can see now that moral perfectionism is premised on self-reflection and self-scrutiny. As we move forward to our discussion of the constant monitoring of our motivations and intentions to make sure they are compatible with a virtuous life, we can foreshadow the conclusion of this discussion, which is not psychologically problematic and neurotic, as the supererogationist supposes. The person who is overfull with moral demands will not necessarily lose hope, as Urmson believes.[17] Instead, it is perhaps ironic, or at least surprising to our assumptions about personal fulfillment, that this moral urgency is the condition of true self-esteem. Continually questioning whether we could do more does not preclude some degree of moral self-satisfaction. While he may look neurotic to those who are morally apathetic, the Kantian moral agent will not be given to despair: indeed, *working for progress is the only way to overcome despair.* Making this point highlights the extreme difference between the moral notion of self-esteem and the popular, empty notion of self-esteem. True self-esteem is not only compatible with, but dependent on, self-scrutiny and self-criticism. Self-esteem is based on self-respect, and self-respect, for Kant, is based on respect for the moral law. As we shall see, moral commitment does not lead to despair and condemnation—a sort of hatred of life and lack of pleasure—as the supererogationist thinks it does. In fact, moral commitment leads to self-understanding, honesty, appreciation of those things that really matter in life, self-esteem, and psychological health.

SELF-SCRUTINY

Psychologically, a great amount of self-understanding and self-scrutiny is required for virtue. Without some assurance that we understand ourselves and that we're being honest with ourselves, we cannot even begin to strive for virtue; if we cannot trust ourselves, we cannot even trust that our ends are actually good at all. For Kant, the supreme condition of moral goodness is a good will; yet a good will is something toward which we must continually strive. Kant is skeptical that anyone can simply know that he or she has a good will; instead, he believes that it is always possible that we choose to conform to the demands of morality merely because it is easy for us or because we have some other cooperating motivation. For this reason, self-improvement requires self-scrutiny.

Kantian moral theory inspires, and requires, self-improvement, which is necessarily based on self-understanding and self-criticism. In the "Doctrine of Virtue" Kant argues that "the first command of all duties to oneself" is "to 'know (scrutinize, fathom) yourself'" (DV 441):

> That is, know your heart—whether it is good or evil, whether the source of your actions is pure or impure… Moral cognition of oneself, which seeks to penetrate into the depths (the abyss) of one's heart which are quite difficult to fathom, is the beginning of all human wisdom. For in the case of a human being, the ultimate wisdom, which consists in the harmony of a human being's will with its final end, requires him first to remove the obstacle within (an evil will actually present in him) and then to develop the original predisposition to a good will within him, which can never be lost. (Only the descent into the hell of self-cognition can pave the way to godliness.) (DV 441)

Oftentimes our motivations are hidden. Especially if we have selfish motivations or feelings that we unconsciously judge to be unacceptable, we try to hide them from ourselves. Other times we resist the simple truths of self-criticism. The "evil will" that Kant refers to is one that resolves to seek personal interests over fulfilling one's moral duties. We often fight against learning that we harbor selfish desires, and being forced to realize that we have them can be painful, a "hell." Plus, we also often harbor repressed negative emotions that can, in this state of denial, inspire problematic behaviors. It is not fun to have negative emotions, just as it is not easy to accept that a choice is morally flawed because doing so requires change, but we have a moral duty to face the truth.

Virtue requires that we adopt the ends of self-perfection and the happiness of others. Setting ends for oneself requires, first and foremost, that one be aware of one's ends as they stand so that one can know if one truly affirms them. In the quote above we see the opposite side of the coin: only when one is truly capable of self-consciously setting ends for oneself, can one strive to take on the fulfillment of duty as one's end. It may seem redundant to argue that one has a moral duty to make oneself able and willing to perform moral duties, but drawing this connection between moral action, intention, and the entirety of one's conscious self-understanding shows us the special, yet natural and purposive, role that moral reason plays in human consciousness. Consciousness must both be oriented toward seeking means of fulfilling our moral duties and also examining our present thoughts and feelings to make sure that they are morally acceptable and conducive to virtue. It would be mistake to represent these tasks as distinct: looking toward our goals and looking at our motivations. Instead, the evaluation of our motivations makes up the environment in which we are able to formulate the ways in which we must perfect ourselves and promote the happiness of others. Self-scrutiny, or, what we might normally call conscientiousness, with the aim of fostering moral commitment, is just what it means to develop self-consciousness since that which we are examining is not entirely in the past, but mostly in the present and related to our actions.[18]

Does Kant's statement that it is our duty to scrutinize ourselves in the effort to cultivate a morally good character lead to neurosis? Does it lead to a person who thinks about morality too much? Kant does not think that this is the case. First of all, as Baron points out, Kant criticizes the moral fanatic, that is, the person who makes amoral choices, such as what color of shirt to wear, into moral choices. Clearly, some things require moral deliberation and others do not.

The problem is that, in this age of global trade and environmental degradation wherein our choices implicate us in thousands of relationships of which we are (intentionally?) unaware, we are confronted with legitimate moral questions everywhere we turn and we cannot possibly address all of them, but we can try, and the first step of this moral effort is awareness—of the consequences of our actions, for example. Striving for this kind of knowledge is not neurotic; it is intelligent. Indeed, it is the denial of these effects and the attempt to focus only on oneself that is closer to the cause of neuroses. In what follows, we will follow this thread of the necessity for conscientiousness (being honest with oneself) to see its psychological fruits as well as the psychological fruits of its opposite, lying to oneself.

SELF-RESPECT AND SELF-ESTEEM

The term "self-esteem" has recently fallen out of favor in popular discourse, and rightly so, because it expresses a hollow idea. Formerly, popular psychology involved the idea that adolescents need positive self-esteem, and now we hear that adolescents are too full of themselves for having been praised without warrant! Aside from the fact that it is unhealthy for one to irrationally hate oneself, there is little content to the idea that one should *like* oneself. Kant, on the other hand, gives this notion its proper content, as he spends a considerable amount of time discussing its proper meaning and, since it must be conditioned by moral worth, its means of achievement.

For Kant positive self-esteem has two meanings, or perhaps we might say, two-levels of ascending meaning. First, positive self-esteem comes from self-respect, or dignity, that one owes to oneself simply because one is human.[19] Kant argues that we have a duty to respect ourselves, just as we have a duty to respect humanity in general. As a feeling, self-respect comes from the comprehension and performance of duties to ourselves: to stand up for oneself and not to harm oneself, for example. This general duty translates into our behaviors in a variety of concrete ways: "Be no man's lackey—Do not let others tread with impunity with your rights—...Do not be a parasite or a flatterer" etc. (DV 436).[20] Self-respect and conscientiousness are related because lying to oneself, by denying the importance of certain emotional needs, for example, in order to avoid difficult conversations or choices, is just as much of a failure to respect a person as lying to another is. In this way, we can see that performing self-respect is not always easy. Nevertheless, we ought to think, "I owe it to myself" (DV 418).

In addition, we should treat our bodies with respect, never as a mere means to getting something else, like money or pleasure. We have the duty to preserve "the dignity of humanity" in ourselves; there is, therefore, "a prohibition against depriving himself of the prerogative of a moral being, that of acting in accordance with principles, that is, inner freedom, and so making himself a plaything of the mere inclinations and hence a thing" (DV 420). We owe this to ourselves no matter what. We are rational beings, and when we treat ourselves—or let other people treat us—like tools or instruments, we debase ourselves. There is nothing inherently wrong with physical labor or physical pleasure, or even bowing to the will of another, but only *as* a *rational* being. We must often treat our bodies and other people's bodies as means to get things, but we must never treat ourselves or other people *merely* as means to get something, overlooking consent, respect, and other human, psychological and moral, needs. Without mutual understanding, which involves the whole of one's personhood, and communication (in sex, for example), there is disrespect and coercion. To settle for less is to debase oneself.[21]

Humans, as rational persons, have inherent value—Kant's word is "dignity"—and deserve respect because they are capable of morally practical reason. It does not matter that a person may have behaved badly in the past; she is still capable of morally practical reason and has inherent value because of it.[22] Self-respect is not something that we owe to ourselves because we are special, nor is our respect for other people contingent on their past behaviors or achievements; the respect that rational beings deserve is unconditional for Kant. When Kant writes that we must recognize a "sublime moral predisposition in ourselves" (DV 432) and that "humanity in his person is the object of respect which he can demand from every other human being, but which he also must not forfeit" (DV 435), he also mentions the absurdity in thinking or acting as though we value more or less than others. Indeed, our self-esteem is based on human equality. We must always see ourselves in this light: in terms of our inner worth and ability to see the truth, not in terms of performance or a comparison with other people.

As we saw in chapter 5, our evaluation of others is tied to our evaluation of ourselves. Kant argues that recognizing that all people have a moral nature and respecting them because of it will "dispel *fanatical* contempt for oneself" and make us realize that humanity itself can never be held contemptible (DV 441). It is important to also realize that a number of vices follow from the fallacy of thinking that moral respect is rewarded for performance instead of as the natural right of every human. Envy, "the propensity to view the well-being of others with distress" comes from using a comparison with others for the standard for our own well-being instead of our own intrinsic worth (DV 458). Malice results from the natural illusion of thinking that we are better off when others are worse off. Furthermore, virtues like beneficence and forgiveness are only so many different expressions of *unconditional* human respect.[23]

This Kantian orientation for self-esteem is important because it rescues us from trying to base our worth on comparisons with other people, on acquiring material goods, or the satisfaction of other inclinations or accomplishment of arbitrary goals.[24] Judging that one is better than someone else in some respect, or accomplishing a certain, arbitrary goal, may make one feel good, but this feeling is destined to be fleeting since we must necessarily question the worth of this other person or of the goal. Unless we tie our self-regard to inherent moral worth, we will discover that it is flimsy.

The first level of self-esteem is a *duty* we *owe* to ourselves.[25] The second level of self-esteem is something that we achieve: we feel good about ourselves when we have done something good, that is, something that has moral worth. When we succeed in behaving morally we *feel* our moral worth and esteem ourselves positively.[26] Self-esteem is a product of moral consciousness and accomplishment.

Kant describes the satisfaction that one receives from moral behavior in a number of different ways. In feeling above the caprice of inclination and deserving of happiness, moral behavior is its own reward.[27] Kant calls this feeling "ethical reward" and describes it thus:

> A receptivity to being rewarded in accordance with laws of virtue: the reward, namely, of a moral pleasure that goes beyond mere contentment with oneself (which can be merely negative) and which is celebrated in the saying that, through consciousness of this pleasure, virtue is its own reward. (DV 391)

We experience joy in acting morally because, as Beck puts it, "reason's interest is being furthered."[28] It is difficult to explain the reason that we feel good when we do good, even if doing so is extremely difficult for us, but this does seem to be the case.[29] In addition, if such is in fact the case, as with all moral feeling it would evince an underlying unity between reason and sensibility.[30]

Proper self-esteem is the ability to be satisfied with oneself and be at peace because one believes that one has accomplished that which is most important—even in a small way. It is like resting after a hard day's work.[31] Of course, this does not mean that we can ever be finished with the demands of morality, rather cultivating a good will ensures that we will be vigilant in pursuing virtue. Without this commitment and desire to do more we are plagued by thoughts of our inadequacy:

> [A] righteous person cannot think himself happy if he is not first conscious of his righteousness; for, with that attitude, the reprimands—which his own way of thinking would compel him to cast upon himself in the case of transgressions—and the moral self-condemnation would rob him of all enjoyment of the agreeableness that his state might otherwise contain. (CprR 116)

Moral guilt stands in the way of self-esteem for those who try to base it on anything other than moral worth.[32] Indeed, both levels of self-esteem are related in that they are both subordinate to a comprehension of moral duty. Self-respect follows from the fact that the moral disposition is itself "sublime" (DV 435). When we conceive of self-esteem in this way, it makes it clear that it is subordinate to morality; indeed, Kant argues that self-esteem contains the feeling of humility within it because it implies a comparison of ourselves with the moral law, in comparison to which one always feels subordinate. This second level of self-esteem is closely related to the self-respect we feel for ourselves because, in fulfilling our duties, we feel our own connection to goodness, which makes us feel that we have inherent worth.[33]

Kant's contrast between happiness, a term that is not necessarily positive for him, and morally worthy happiness (what he calls the "Highest Good") can help us to further understand the true meaning of self-esteem.[34] Kant faults the Ancients (the Stoics and the Epicureans) for overlapping the concepts of goodness and happiness, believing either that goodness makes one happy or that happiness is itself the good. Kant remarks that, unfortunately, neither is the case. In other words, there is absolutely no guarantee that (materially) good things will happen to (morally) good people.[35]

Instead, he makes a distinction between self-satisfaction and happiness, the latter of which he believes is only a product of the fulfillment of the inclinations, if such fulfillment is even possible.[36] (Kant argues that inclinations vary and the pursuit of their total satisfaction always only creates "an even greater void than one had meant to fill" (CprR 118).) Self-satisfaction, on the other hand, is a true respite from the ping-ponging of attention from one pleasure to the next. Self-satisfaction is based on proper moral understanding and estimation—the understanding that inclination is not as important as morality. Self-satisfaction, which is the same as what I am calling the second level of self-esteem, is the feeling of deserving happiness.[37] It requires that we distinguish between the pleasing and the disagreeable, on the one hand, and good and evil, on the other.[38] The second is an objective, moral evaluation that is not dependent on the amount of pleasure something promises to give.

Kant argues that it is easy to get caught up in seeking our own happiness, as it is easy to think that "*everything* hinges on our happiness."[39] He believes that we naturally attend to our happiness, but that it is morally necessary that we also judge as moral beings and make moral worth the supreme condition of our happiness. This conditionality means three things: (1) that we limit our practical maxims when they conflict with the moral law, making it our meta-maxim to always follow the moral law above all others; (2) that we cultivate morally worthy practical maxims as our indirect duties require; and (3) that we judge happiness to be morally good only when we have also achieved moral worth. This last meaning is the highest definition of self-esteem: we can and should be pleased with ourselves only when we have achieved some degree of moral worth. Moral worth comes from having a good will; a good will is one that is determined above all by the moral law.

PSYCHOLOGICAL HEALTH AND CONSCIENTIOUSNESS

Although I agree with Kant that we do not care about morality because we are trying to get something out of it for ourselves, namely happiness, the fact that I believe that we are naturally prone to moral thought (however immature) leads me to the further conclusion that developing this moral thinking leads to some degree of personal fulfillment and psychological health. Similarly, ignoring morality leads to psychological disorder. In taking morality seriously, one is inspired to scrutinize oneself in order to better perfect oneself and promote the happiness of others. Self-understanding, self-respect, and self-satisfaction are, perhaps, general prerequisites for emotional self-improvement as they stand in for general psychological well-being.

Kantian moral theory has been described as cognitivist, universalist, and formalist.[40] Cognitivism is the position that moral statements have truth conditions. Noncognitivists hold that moral judgments are forms of noncognitive self-expression, such as desires or statements of approval or disapproval. We can see immediately that moral noncognitivism assumes a noncognitive theory of emotion; statements of approval or disapproval are taken to be noncognitive by the noncognitivists and not susceptible to rational justification. Kant, on the other hand, encourages us to evaluate and be responsible for our emotions and desires.

Cognitivist moral theories imply that moral judgments are things about which we should *do* something. If a moral judgment is not really a judgment at all, but merely an expression of my own personal taste, as the expressivist holds, then my judgment that, for example, dog-fighting is wrong does not compel me to enter into a discussion with you about dog-fighting if you disagree with me.[41] I can simply say: to each his own. If I see dogs being fought then, even though I *feel* that it is wrong, I need not be compelled to stop it. I would just think: "Oh, they might not think it's wrong; I should just leave so I will not have to be confronted with my feeling of its wrongness." Similarly, many of us are told that morality refers to our personal values and commitments. Realizing that it would be narcissistic to be committed to something purely personal, this position is also the injunction *not to act* on our moral thoughts and feelings, not to *judge* others.[42] According to this theory, it would be *culturally insensitive* for me to object to a practice like dog-fighting simply because some other people live that way.[43] It is this moral denial, or moral weakness or cowardliness—you will support these strong terms if you agree with moral realism—that is a genuine moral, and now psychological, problem.

For all people the demands of morality are experienced as compulsions, yet following through on these moral inclinations poses a challenge. Hence moral relativism, or backing down from defending one's beliefs, is an equally natural stance. Nevertheless, the latter position results not from expressing, but from denying one's moral feelings. Moral subjectivism, even though it seems to affirm the emotions, really talks down to them and fails to take up their inherent challenge of self-improvement. In the case of emotions, which call out for an enquiry about the best mode of action, the emotivist, relativist turn yields nothing but a dead end.

As we have seen, Kantian moral theory encourages self-scrutiny. Kant's notion of practical reason also includes the idea that scrutinizing one's motivations better creates a harmoniously integrated self, one in which conscious motivations and unconscious motivations and preconscious thoughts match up. Moral deliberation requires that we reflect on our moral feelings to discover and evaluate the reasons that are at their base. There is no reason to think that emotional reasons are more likely to be immoral or irrational than our more explicit inclinations. In fact, if the moral sense theorists are right, uncovering the reasons beneath our emotions, and more effectively acting on them, will help us to be better people. A cognitivist moral theory requires that we uncover the reasons implicit in our emotions and morally evaluate them. This emotional engagement creates a more harmoniously unified self.[44]

In addition to accepting our emotions and taking them seriously, moral cognitivism encourages us to act on our legitimate moral feelings. Often our moral decisions are based on mere mood or whim, not on my considered moral convictions, but this random, flippant decision-making corresponds with an erratic, fractured identity. For example, I personally believe that I should offer people without means of transportation, for example, hitchhikers and poor people, a ride in my car, provided I do not believe that doing so will endanger me. Although, I have fairly well-developed reasons for this conviction, I hardly ever act on it. My actions in this case, either to offer a ride or not to offer a ride, are usually based on mood and inclination. To be fair, I have reasons for my failures in every case, for example, I think that my husband might disapprove, I do not want to be bothered with talking to someone, I want to do something else instead and without delay, and so on. Still, I do not honestly believe that any of these feelings or reasons outweigh the importance of helping others. When I do give a needy person a ride, it is because my affects support that outcome, for example, I am feeling particularly leisurely because it is a sunny day and I am not in a rush or I am feeling particularly loving and confident because some of the circumstances in my life support this mood.[45]

It may seem that someone who unapologetically embraces this whim-based way of deciding in each case is more "in touch" with her emotions. My argument is the contrary: that if I were in fact to act on principle more often, or all the time, my behavior would be more emotionally aware. I implicitly value my cognitive (and moral) reasons more than my fleeting reasons for acting. I feel that the cognitive reasons better represent my ideal (or true) self and my failures to live up to my ideal are manifestations of cowardice. When I act on principle, I am pleased with myself, and I perceive my failure to so act as a personal flaw.[46]

A cognitivist moral theory promotes emotional self-improvement in a third way also by encouraging us to formulate and reevaluate our intellectual principles. Inclinations, even though Kant defines them as habituated desires, are often more mutable and arbitrary than our more considered principles. When we act on these unstable inclinations, we do not learn anything about ourselves, or, rather, anything about our beliefs. We may discover that we are much more likely to listen to an upset friend after we have had a cup of coffee in the morning, but we cannot discover if we *should* have the coffee in the future, if listening to the friend is something *good* to do. We may discover that we like pleasing people by giving them compliments, but without an understanding of the principle, or moral reasoning, behind this action, we cannot test whether or not this pleasure is something we ultimately value, something we should strive for, and if it is better or worse than any other kind of pleasure. On the other hand, when we assert a conviction and act on it, we are then in a position to learn from our action. Taking morality seriously is, in a sense then, taking ourselves seriously.

Emotional self-improvement appears to require that we act with conviction, which is a part of having courage and a strengthened sense of self. We can see this in the fact that the failures to know what one wants and to act on this knowledge are related to codependent relationships and general malaise. The ability to make choices and stand by those choices, because we believe them to be better than the opposite, is a large part of the ability to engage in emotional commitments, respecting ourselves and others. These commitments are necessary for and expressions of psychological health but are also themselves, in turn, often morally required.

Some might object that Kant's vision of human nature is precisely one that is psychologically unharmonious, with purified reason in constant tension with our selfish passions. Just as it is obvious that we will always have emotions that we need to experience and think through, Kant is convinced that we will always have inclinations that practical reason will need to evaluate and adjust. Kant portrays this process as one of struggle, but we will not see this as offensive if we keep in mind that it is often, in fact, the case that confronting one's selfish tendencies or resistance to acknowledging certain emotions is, in fact, a struggle. It is necessary to recognize the emotional reactions and difficulties that are an intimate part of the human condition. Of course, our inclinations are not always selfish, as Kant would sometimes have us think, and our emotions are, of course, not always based on flawed thinking but are often moral and rational, but still inclinations and emotions need to be evaluated and integrated with conscious reason and conscience. Without this integration, our emotions are blind and our cognition empty, or, at least, deficient.

Allison discusses a sense in which the work of practical reason is psychologically integrating: he likens practical reason's ability to unify inclinations under a maxim to speculative reason's transcendental unity of apperception, which unifies all experiences under one consciousness (the "I think").[47] Practical reason decides which inclinations it will take as motivating forces. As Sullivan describes it:

> To allow oneself to be ruled by freedom-destroying inclinations is the essence of vice. Pure practical reason thereby requires that we test our maxims to make sure that they are permissible. Freedom requires us to bring all our capacities and inclinations under the rule of reason, but to do so calmly so as not to rely unwittingly on inclinations for motivation.[48]

Nevertheless, Kant holds that we *freely* choose against morality when we lie to ourselves, so perhaps a better way to describe that which opposes virtue would be as Grenberg does, drawing from Kant's *Religion* essay, as a sort of weakness of reason.[49]

Unfortunately, the idea that morality and psychological health are related can be misunderstood in a way that erases the demands of morality. There is a difference between mere psychological coherence and moral rectitude. Many recent moral theorists, like Velleman, argue that immorality is the same as psychological incoherence.[50] Gewirth argues that to deny others the right to freedom and well-being is to deny our own such rights and hence ultimately threatening to the self.[51]

Working in the Kantian tradition, some theorists interested in the notion of autonomy have developed it into the more robust idea of reflectively endorsing one's volitions, in the sense of discovering and acting on one's "true" desires. Frankfurt has an "authenticity" interpretation of autonomy, holding that autonomy requires that one's second-order desires identify with one's first-order desires.[52] Wolf characterizes Frankfurt's theory of freedom as a "Real Self Theory" because he believes, along the same lines as Kant's notion of autonomy, that an action is free if it issues from the true self.[53] The self of pure reason is our most true self, for Kant, but for Frankfurt the true self is more immanent in our desires. We may or may not identify with our desires, but those desires that best express the self are those with which we are most involved, to the point of having what he calls "volitional necessity."[54]

Both the means of and motivation for these types of arguments are strange to me. Nevertheless, the tendency to break Kant's equation between autonomy, pure reason, and the categorical imperative is not uncommon. Whether or not these attempts to prove that self-coherence and moral worth are the same thing are successful, they overlook the necessary, objective content of moral reasoning, that is, the categorical imperative. At the level of morality, reason unifies inclination according to the thought of pure lawfulness, or universality.

Virtue also requires that we adopt *virtuous* maxims, namely to perfect ourselves and promote the happiness of others. Morality, therefore, requires a higher level of reflection. Not only do we consciously decide what we are going to do, as Allison suggests, but we question whether or not it is really the right thing to do. Moral deliberation requires that we reflect on our emotions, discover their cognitive bases, and then try to square those cognitions with the moral law. My argument is that this further process of *moral* deliberation has the effect of promoting psychological health.

It is true that desires and emotions often contain or accompany self-referential evaluations, but it seems mistaken to suggest that goals are important *because* we care about them. They are important *to us* because we care about them, but that conclusion is tautological and we need to ask the question "should this be important to me?" Herman argues that we ought to be critical of our desires and ends, especially in our intimate relationships, but we should not think that moral deliberation must take us away from these ends. Instead, it is a part of them and enhances them.[55] On the other hand, sympathy and caring can sometimes be false moral impulses: we also need to evaluate our emotions with an eye to their effectiveness in promoting good outcomes. In some ways, as Solomon polemically argues, emotions are acts, and as such, they themselves should be morally evaluated.[56] Our feelings about our emotions (or higher-level beliefs) are a good place to start when seeking to evaluate our emotions, but moral evaluation must go farther. It is necessary to bring in higher standards, for example, universality and respect, in order to more fully scrutinize and evaluate ourselves.

Someone might object that in refusing certain inclinations, we are not unifying inclination and reason at all, but repressing one, yielding nothing but internal strife. Michael Stocker suggests, with his notion of moral schizophrenia, that it is psychologically unhealthy to have desires and yet try to prevent oneself from acting on them. Nevertheless, to think that we could ever have an easy and harmonious integration between our desires and actions is naïve. In terms of immediate and long-term subjective experience, there is a lot going on, to put it crudely. We are never in a position of totally understanding ourselves, and so, we are always in a state of schizophrenia (if we are willing to play with the name of a serious mental disorder in that way).

Munzel argues that since Kant believes that we have a natural consciousness of our moral capacity and that his theory of human nature shows that it is responsive to moral direction and does not need to be dominated.[57] Nevertheless, admitting that the emotions are not irrational means that they too must play a role in moral deliberation, although not as blind intuitions, but as the markers of well-reflected reasons. We might liken this back and forth characteristic of moral deliberation to Rawls's notion of reflective equilibrium.

The retort to Stocker and other similar objections lies in the conviction that consciousness of duty, or compulsion under the moral law, is a "fact of reason" and that morality is a natural human calling. Kant calls the moral law a "fact of reason." In other words, we are conscious of moral constraint. At the most foundational level of his argument Kant appeals to common sentiment.[58] People do, in fact, feel the authority of morality (a psychological examination of the problems associated with denying one's moral feelings demonstrates this). We might amorally reflect on our emotions and integrate them into our pragmatic, long-term goals, but there is no guarantee that we will feel any real identification with this integrated consciousness. There is no value, or even sense of self, in merely having a coherent identity. On the other hand, there is a sense in which Kant's notion of duty promises to uncover our *truly* true selves, or our *better* selves. The "purity" of moral reason refers not merely to the exclusion of empirical determination, but also to its superlativeness.[59] Our moral/rational consciousness is also the truth of our emotions.

Indeed, moral commitment gives life meaning, and, without it, it is difficult to imagine why one would be satisfied with life. Just as Kant contrasts the good will, as that which is good in itself, to things like health or wealth, which can be misused, in true Aristotelian fashion, Kant then argues that the final purpose of existence "is a purpose that requires no other purpose as a condition of its possibility" (CJ §84). Kant implies that "unconditioned legislation regarding purposes" constitutes the purpose without a purpose or the unconditioned condition. In other words, in some respect, our highest purpose in life is no purpose at all: we do not make our moral choices in order to get something out of them, we choose based on what is right and wrong. The action itself is its own purpose.

Overall, the argument of this chapter is that the person who lives with moral commitment lives a life of purpose. Although she is not immune from bad luck, she lives a good life, not just in the moral sense of the term. Nevertheless, it seems that the fear of moral commitment leads people to run from it with a cure that is worse than the illness, so to speak. Imagine for a moment a person trying to achieve relaxation and happiness directly—not as moral desert—by taking a yoga class, for example. In order to fit yoga into her busy schedule, she rises early, has a quick, microwave breakfast, hops in her car to drive to yoga, picks up a latté on the way, then participates in the class, meditation and all, wherein she is asked to clear her mind. Imagine the difficulty involved in quieting this mind that has likely ignored so many thoughts in its quest for peace and convenience: Did the emissions from my car cause the largest typhoon ever to hit the Philippines, killing thousands of people? Will the lid to my coffee cup and the packaging from my breakfast go on to kill multiple animals and people? Was the coffee I drank grown by people working in near-slavery conditions? Did people cut down the rain forest in order to grow it? Did shipping it here to me also cause the typhoon? Do the people who made it and my breakfast for me make enough money to feed themselves and their children? Can they go to the doctor? Did the animals I ate for breakfast live a filthy life of intense suffering? Are the workers who slaughtered them slowly becoming sociopaths? . . . It would be understandable if this person has trouble quieting her mind. Although most people do not ask themselves these questions, no rational human being is capable of living "in the moment" enough to completely succeed in not wondering about the consequences of her actions. Perhaps it is not as easy to ignore the moral dimension of one's actions, that is, avoid the questions "Where did it come from?" "Where will it go?" and "What if everyone did that?" as it at first appears. Compare this meditator to the person who allows herself to attempt to think through all of these moral demands and is committed to trying to make the world a better place—the person who is committed to self-improvement, but knows that virtue is "always in *progress* and yet always starts *from the beginning*" (DV 409). The latter person has much more work to do and an uphill battle to fight, but, I ask, who is more likely to find peace?

8

A Morally Informed Theory of Emotional Intelligence

The goal of this chapter is to apply the work that we have done thus far to the psychological notion of emotional intelligence. No doubt my reader has already heard the term "emotional intelligence" and has some associations brought to mind by it, largely due to the prevalence of media attention given to the idea in recent years. Some psychologists are positively giddy at the amount of popular attention their field is receiving; others are suspicious and denounce the whole notion as "pop psychology." Here I review the research being done on emotional intelligence in order to clarify the concept. We shall see that there is a great deal of disagreement between psychologists over the meaning of the term. Some accounts stand out for their accuracy, while others, like Daniel Goleman's book *Emotional Intelligence* (which led to the popularity of the term itself), are problematic. Definitions that focus on the automatic nature of emotions prove unable to explain the means by which people might become more emotionally intelligent, as does identifying or assimilating emotional intelligence with cognitive intelligence. The focus on emotionally intelligent behavior, on the other hand, which we see in the work of those who focus on the practical question of teaching emotional intelligence, resists positing a latent, innate ability and instead focuses on achieving psychological health. I argue that the idea of "emotional intelligence," as well as "emotionally intelligent behavior" and "emotional literacy," can serve as a basis for self-improvement and psychological health as long as we resist positing a latent, innate emotional ability and instead focus on the same goals of emotional self-understanding and moral commitment that we have been discussing thus far.

Based on the work accomplished by this book so far, it could turn out that moral inquiry lies at the heart of emotional intelligence and that emotional intelligence is inherently a moral concept. In fact, current work on emotional intelligence dovetails with moral inquiry in many places, as with school programs that promote conflict resolution and prevent bullying. Without moral guidance, it is impossible to delineate emotionally intelligent behaviors from those that are emotionally coercive or sociopathic. Unfortunately, psychologists, being scientists, are often not comfortable navigating moral normativity, and hence the psychological construct of emotional intelligence remains underdeveloped in this regard. Moral theory and a philosophical notion of well-being can and must guide our search for emotional intelligence. It is the job of a philosopher to pick up where the psychologists leave off.

Furthermore, the notion of emotional intelligence can inform moral theory. As we shall see, since emotional intelligence involves the health and happiness of the whole person and her ability to make good decisions, engage in healthy relationships, and prevent and resolve conflicts, it makes little sense to command moral behavior without commanding, and facilitating, emotional intelligence. In other words, emotional understanding (including emotional literacy), evaluation, commitment, and conflict resolution skills are not just good "social skills"; as social beings, the urgency to develop these skills is a moral one.

Differing Definitions of Emotional Intelligence

Because there are so many different definitions of emotional intelligence, it is safe to assume that two people in two different places—two different universities or offices, for example—do not mean the same thing when they both use the term. Owing to its popularity, there are over sixteen different tests used to measure emotional intelligence or something like it for the sake of research, education, or profit, and there is little to no evidence that their results correlate.[1] Nevertheless, there are three prominent definitions of emotional intelligence: those by Salovey and Mayer, Bar-On, and Goleman.

Salovey and Mayer originally defined emotional intelligence as "the subset of social intelligence that involves the ability to monitor one's own and other's feelings and emotions, to discriminate among them and to use this information to guide one's thinking and actions."[2] By "a subset of social intelligence," they mean to refer to Thorndike's notion of social intelligence, "the ability to understand and manage people," and thereby to tap into the history of work in intelligence studies.[3] Salovey and Mayer's definition is often broken up into its constituent parts, so that emotional intelligence is said to involve four distinct abilities: the ability to perceive and appraise emotion, the ability to use emotion to facilitate thought, the ability to understand and communicate emotion concepts, and the ability to manage emotions in oneself and others.[4] Even within this one camp, we can find one research team having subjects listen to vocal intonations and another research team looking at something as different as the ways that students prepare for tests, both expecting to draw conclusions about "emotional intelligence."

The models of emotional intelligence developed by Bar-On and Goleman are both characterized as "mixed" rather than as a "pure" ability-based model. Bar-On developed the emotional quotient inventory (EQ-i), which is based on 15 subscales and aims to predict the degree to which an individual interacts with her environment in such a way as to promote her own psychological well-being. This model defines emotional intelligence as "an array of noncognitive capabilities, competencies, and skills that influence one's ability to succeed in coping with environmental demands and pressures."[5] Describing each of these subscales would prove tedious, so I will merely name them: self-regard, self-awareness, assertiveness, independence, self-actualization, empathy, social responsibility, interpersonal responsibility, stress tolerance, impulse control, reality testing, flexibility, problem solving, optimism, and happiness.[6]

Most people outside the field of psychology associate the term "emotional intelligence" exclusively with Goleman's book. Goleman is known for his striking and sweeping claims, such as the claim that emotional intelligence is more important than IQ in determining life success. Working with Boyatzis and Rhee, Goleman has developed a model that includes twenty-five different areas of competency.[7] He refers to emotional intelligence in a variety of ways: "self-control, zeal, persistence...the ability to motivate oneself"; "to reign in emotional impulse; to read another's innermost feelings; to handle relationships smoothly"; "to persist in the face of frustrations, to control impulse and delay gratification, to regulate one's moods and keep distress from swamping the ability to think; to empathize and to hope."[8] He often relates it to life success, describing it as a "meta-ability, determining how well we can use whatever other skills we have, including raw intellect"; he sometimes calls it "people skills."[9] Goleman's positive description of emotional intelligence centers on the increased learning potential that comes with something we might normally call "having a good attitude." He paints the picture of the contagiously good mood of a bus driver as an example of positive emotions working to better our lives.

The bulk of Goleman's discussion is dedicated to convincing his reader that emotional intelligence is an important quality; he does this by illustrating the problems that a lack of emotional intelligence supposedly causes. Goleman describes those who lack emotional intelligence as "those who are at the mercy of impulse—who lack self-control."[10] Seemingly taking anger as the prime example of emotion, he posits that those without emotional intelligence are those who "lose it" or are subject to an "emotional hijacking."[11] His notion of a lack of emotional intelligence is even broader than his notion of emotional intelligence, and includes marital discord, insensitivity and general meanness, stress, abuse, trauma, and homicide.

Goleman has worked with others to develop the "Emotional Competence Inventory" (ECI). The ECI measures four dimensions of emotional competence: self-awareness, social awareness, self-management, and social skills. Conte and Dean report that "no empirical, peer-reviewed journal articles are presented to support the validity of the ECI."[12] They also conclude that there is little evidence that the ECI is able to discriminate reliably between people or to predict socially relevant outcomes. Despite such criticism, other theorists, like Salovey and Sluyter, rely on Goleman's popularity to increase the impact of their work, even while rejecting his understanding of emotional intelligence.[13]

Although there is disagreement over the definition of emotional intelligence, the idea itself is in many ways nothing new. As Murphy and Sideman note, "EI is often seen as a new name for constructs that have been studied (sometimes with limited success) for decades."[14] Most new work on emotional intelligence realizes this fact and attempts to connect emotional intelligence to traditional areas of research, such as stress, addiction, family and marriage relationships, child development, or personality. The notion of emotional intelligence seems to draw from the basic idea behind clinical psychology: the idea that people can improve their lives by gaining insight into their behaviors and motivations. Perhaps the best way to understand the notion of emotional intelligence is to consider the reasons it is so attractive. We like to think that we have a substantial amount of control over our lives and that we can improve ourselves. The notion of IQ tends to leave people with resentment; EQ, on the other hand, promises to be more egalitarian.

Mixed models of emotional intelligence have been shown to overlap with measures of personality.[15] Goleman's ECI, for example, overlaps with traits measured by personality tests such as conscientiousness, emotional stability, extraversion, and openness.[16] This is a problem only because those who study personality take it to refer to relatively fixed traits; whereas those who study emotional intelligence describe it as something that can be inculcated and altered. It is possible that overlap between notions of emotional intelligence and personality is not a problem, if we understand personality differently. It is strange to me to suggest that the "Big Five" personality traits are value-neutral. Neuroticism, for example, has been shown to correlate with negative life-outcomes. Openness, on the other hand, seems to be the opposite of closed-mindedness, which seems like necessarily a bad thing—bigotry?[17] While it is probably the case that people usually maintain basic personality traits throughout their lives, we do not have to conclude thereby that it is impossible to change your "personality." It seems reasonable to try to help people who are neurotic or closed-minded become more relaxed and open-minded, even if it is very difficult for them to change. Most psychological messages we get from popular culture are doing exactly that, and it would be remarkable if personality psychologists could prove that this kind of change is impossible.

In a somewhat similar vein, Goleman suggests that emotional intelligence might also be called "character." It is interesting to consider the reasons for the spate of popular attention to the notion and importance of character. Attention to character, as with emotional intelligence, can be useful when it is used to draw attention to the full context of a person's life and upbringing. On the other hand, we should have serious concerns about the notion when it is used, for example, to draw attention to the innate qualities of children and away from the responsibilities of their parents and communities.[18] The idea of teaching emotional intelligence (as well as character) in school is compelling, if we can find the right way to do it. On the other hand, teachers and parents should not use the notion of character (or emotional intelligence) as a way to let themselves off the hook for doing what is precisely their jobs, namely to teach. It makes sense that Goleman would want to promise his readers that their destiny is not controlled by IQ, that they can improve, and, at the same time, that there is something special about them, their character, that promises to win them success over their peers.

Most theorists who are currently working with the construct of emotional intelligence agree that it should be thought of as something that is corrigible, and it does seem that a good definition of emotional intelligence will not fall into the trap of thinking that emotional intelligence, like traditional notions of intelligence, is an innate and inalterable quality of a person. Instead it will focus on the teachable knowledge that allows for creating emotionally intelligent behaviors.

EMOTIONAL INTELLIGENCE AND COGNITIVE INTELLIGENCE

The word "intelligence" seems to connote innateness to some. Although IQ tests were first designed by Spearman to measure educational achievement, intelligence has come to be thought of as an inherited trait.[19] Those who study intelligence ("g") largely agree that it is genetically determined: twin and adoptive sibling studies have demonstrated heritability.[20] "Intelligence" is defined as an "ability," and likewise models of emotional intelligence that posit it as an "ability" imply that it is innate. Salovey and Mayer's MSCEIT correlates modestly with IQ, as they believe it should.[21] They write, in the tradition of intelligence studies, that a "new" intelligence ought to correlate with g to a moderate degree: "no correlation at all could suggest the new 'intelligence' is so different that it is not an intelligence at all," and a high degree of correlation suggests that the new intelligence is not new.[22]

The field of emotional intelligence has a troubled relationship with intelligence studies. Coming out of Gardner's theory of multiple intelligences, it similarly promises a way to both challenge and include people in the hegemonic notion of IQ.[23] Goleman clearly hopes to show that EIQ can inherit the glory of IQ. Like theories of cognitive intelligence, Goleman's notion of emotional intelligence claims to predict success in the economic sense of the term. Although he stresses that emotional intelligence depends on our upbringing, defining it by its relationship to IQ has the effect of inadvertently making it seem innate.[24] Goleman falls into this trap himself, referring the emotional intelligence differences among four-year-olds, although surely a four-year-old is still learning how to understand and express her emotions.[25] Goleman makes clear that he intends his notion of emotional intelligence to inherit the traditional notion of intelligence when he writes:

> At age four, how children do on this test of delayed gratification is twice as powerful a predictor of what their SAT scores will be as is IQ at age four... This suggests that the ability to delay gratification contributes powerfully to intellectual potential quite apart from IQ itself. (Poor impulse control in childhood is also a powerful predictor of later delinquency, again more so than IQ.)[26]

Goleman seems to posit emotional intelligence as a special gift or talent that predisposes certain children toward life success and others toward failure.

Many who study general intelligence believe that it is highly stable over time and that attempts to inculcate it have little long-term effect, yet all who study emotional intelligence agree that it can and should be taught in public schools. Salovey and Mayer seem to be conflicted: they want to draw from intelligence models, but they insist that EI ought to be learnable, even when their intelligence-derived notions turn out not to support such conclusions.[27] Caruso, Bienn, and Kornacki doubt that the MSCEIT model allows for corrigibility; yet Salovey and Mayer do not seem to recognize this consequence of their association with IQ.[28] Furthermore, Salovey and Mayer have yet to prove that the "ability" they measure actually translates into concrete behaviors.[29]

There is no evidence that any definition of emotional intelligence is capable of rivaling IQ as a construct that refers to the separate mental system of emotion, especially since there is no evidence that an entirely separate mental system of emotional functioning exists in the first place.[30] Perhaps there is instead some internal relationship between cognitive and emotional intelligence. Brody argues that it makes sense to think that spatial visualization aids in working with emotional information just as it has been shown to help in paragraph comprehension, and that "emotional intelligence is likely to be one component of g, not a substitute for g."[31] It is also possible that some minimal degree of intelligence is necessary for emotional intelligence. Another possibility is that general intelligence aids in emotional intelligence, because emotions tend to be complex and understanding them can be just as difficult. On the other hand, it is also possible that less intelligent people have less complex emotions, and therefore have no more or less difficulty understanding them than do more intelligent people; although, it seems that they would necessarily have more trouble understanding the more complicated emotions of others. At any rate, it is not clear that there is any real similarity between the two notions aside from the term.

Dropping the name of "emotional intelligence" and using terms such as "emotional literacy" or "emotionally intelligent behavior" proves to be a more promising route.[32] Work on the latter topics excels in its careful self-distancing from the tradition of intelligence studies. For example, Ciarrochi et al. describe emotional intelligence not as excelling, but rather as achieving normalcy or health in a world in which

> 33% of people have a diagnosable mental disorder and 50% of us seriously contemplate suicide at some point in our lives... [and in which] we have developed increasingly inventive ways to wage war and kill one another.[33]

They define emotional intelligence in terms of peace and happiness, as well as in terms of the emotional needs and skills that make these possible.

Others also prefer the word "competency" or "literacy" to "intelligence" in order to focus on learning. Brackett and Katulak discuss "emotional literacy" and argue that having an emotional vocabulary enhances one's ability to think about emotions and engage in emotionally intelligent behavior.[34] They also mean to call to mind something like "emotional fluency," or a kind of comfort and ease with having emotions.[35] Saarni and Buckley have developed the notion of "emotional competence" in order to focus on learnable skills: "The skills of emotional competence are learned; their acquisition is influenced by family, peers, school, media, societal scripts, and folk theories of how emotion 'works.'"[36] Furthermore, they eschew the tendency to think in terms of rational mastery and stable dispositions by emphasizing the fact that emotions always take place in the context of a fluid relationship:

> [H]ow our emotional functioning develops in a social context, how it is revealed in our everyday life, depends on the ongoing exchange between a person and her environment. Individual factors, such a cognitive development and temperament, do indeed influence the development of emotional competencies. Yet, skills of emotional competence are also influenced by past social experiences, and learning, including an individual's relationship history, as well as the system of beliefs and values in which the person lives. Thus, we actively create our emotion experience, through the combined influence of our cognitive developmental structures and our social exposure to emotion discourse. Through this process we learn what it means to feel something and to do something about it.[37]

I could not agree more. This focus on one's developmental history and ongoing relationships, as well as values and exposure to emotion—and moral—discourse, helps us to get at the heart of emotional intelligence, and to that which is involved and at stake in developing and maintaining it.[38]

The fact that I am not working in the field of psychology means that I am afforded some leniency not allowed those pushing to define and measure this new construct, and I can question whether or not emotional intelligence is the kind of thing that should be measured in this way. If a test is designed to measure emotional intelligence "reliably," for example, it will give people a score that will remain constant across various situations and over time. To push for a "reliable" measure seems like the exact way that emotional intelligence should not be understood. Following Saarni and Buckley's notion of emotionally intelligent behavior, with its emphasis on the idiosyncrasies of one's developmental history as well as the ways that we continually learn emotional scripts from our ongoing relationships, it would be better to design a test that would seek out problem areas rather than posit a general score. Such a test might ask a person about his or her recent and past experiences with anger, or any other emotion, and whether or not he or she believes that she has handled these situations well, looking at the situation and others involved, as well as moral notions like self-respect and fairness. We can see that, for such a test, measuring and changing emotional intelligence would be internally related, as the subject reflects on her experiences and imagines different ways of understanding them. There is some reason to think that only more contextually sensitive, discursive measures would be possible, if we are even certain that the reification of measuring, as opposed to more forward-looking educative approaches, is useful.[39]

Ciarrochi believes that emotional intelligence relates to life success, but he clearly means something different by "success" than Goleman does. There is, in fact, no support for Goleman's claim that emotional intelligence relates to job performance.[40] We should not confuse emotional intelligence with salesmanship or customer service skills. Nor should we see it as the tendency to fake positive emotions.[41] Emotional intelligence might correlate with job success in jobs that require emotional labor, such as the helping professions, like social work or education.[42] It is also quite likely that a morally informed notion of emotional intelligence will correlate negatively with job success in many careers, since many occupations demand immoral and psychologically unhealthy behavior from their employees. Only a morally informed notion of emotional intelligence can adequately discriminate between different notions of "success."

INTELLIGENT EMOTIONS

Any theory of emotional intelligence must be based on an accurate theory of emotion, and Kant's theory of emotion—as bodily sensations that accompany certain thoughts—helps us to see that sometimes emotions themselves are intelligent. If the emotions are seen to be merely affective reflexes, then there is no possibility of internally merging emotion and intelligence; we are left only with the possibility of stoically "controlling" our problematic emotions, which is theoretically and psychologically untenable. Similarly, we cannot exclude the realm of desire and motivation when understanding an emotion; emotions often represent certain wants and needs, some of which ought to be met. The idea of emotional intelligence, if it is taken out of the context of experience, might seem to imply that we can simply talk through and talk away all of our emotions.

Goleman's theory of emotional intelligence replays the popular stereotypes of and biases against emotionality; this is likely the very reason that it has achieved such widespread popularity. Goleman believes that emotions come from and represent a distinctly different part of the mind than does rational thought. This view is what Jagger calls "The Dumb View" because it assumes that emotions are, in themselves, unintelligent.[43] Goleman employs the popular clichés of "head" and "heart" to refer to "the rational mind" and "the emotional mind." In an appendix devoted to explaining his theory of emotion, Goleman posits that there are two types of emotional responses: those in which thoughts come first and those in which physiological responses come first. He devotes the most attention to the latter and gives dramatic performance, as with the tears brought on by an actress on stage, as an example of the former. (We saw this same distinction and theoretical privileging in Prinz's account in chapter 2.) Goleman's main argument is that the rational mind can be "hijacked" by the emotional mind and that the rational mind must fight back to subdue these irrational forces.

Goleman's theory of emotion offers a glimpse into the world of emotional disorder. The sort of person who has an explosive temper and reacts by attempting to then repress his anger may be "controlling" the emotion, but he is not making progress in understanding it nor being emotionally intelligent.[44] He has not found a long-term, sustainable solution to the problem. We should not follow Goleman in his talk of "controlling" and "mastering" emotions. In addition to overlooking the important and positive role that emotions play in our lives, this approach precludes the intuitive likelihood that emotions themselves contain rational content, that is, that they might themselves be intelligent. Instead, emotional intelligence should be cast as understanding, not controlling, emotions. Talk of "understanding" underscores the ways that reason and the emotions are related; talk of "mastering" reinforces the unfounded idea that emotions are always irrational.

Perhaps it would be instructive to briefly compare Goleman to Spinoza. Like Goleman, Spinoza holds that the task for emotional self-improvement is the overcoming of those emotions that "hinder the mind from understanding."[45] Spinoza, like many in the history of philosophy, understands thinking as an activity and feeling as a passivity. In his *Ethics* Spinoza defines passivity and passion along the following lines: "I say that we suffer when anything is done within us, or when anything follows from our nature of which we are not the cause except partially."[46] The challenge for us now is to transform the prejudice that we have inherited from language and the history of philosophy and understand feeling and thinking as both active and passive. We must identity ourselves with our feelings the same way we identify ourselves with our thoughts. If we realize that certain bodily sensations actually come from thoughts, they will not seem so foreign. Plus, we do not choose our thoughts, nor do we accept all of our thoughts wholesale as always correct or desirable. Similarly, we need not accept all of our emotions. In truth, both thought and emotion are passive and active in the sense that they are part of our process of understanding and acting on our environment. Much about the environment, including our own nature and the nature of reason and morality, is out of our personal control. Nevertheless, as Spinoza would argue, the extent that we can become aware of nature, especially our own moral and rational nature, and act in accordance with it, we are free. Kant's notion of autonomy mirrors this sense of freedom. We often have emotions that express our true (rational and emotional) nature, and when we better understand, think through, and express our emotions, we can come to have even more good emotions.

Although Salovey and Mayer believe that there is a distinct affective subsystem of the mind, they hold that "definitions of emotional intelligence should in some way connect emotions with intelligence," and they describe emotional intelligence as a combination of "the ideas that emotion makes thinking more intelligent and that one thinks intelligently about emotions."[47] It is unclear whether or not they wish to consistently maintain that the emotions in themselves ought to be conceived of as distinct from cognition and motivation, since their new definition remains vague on the key point of the theory of emotion. While Kant is not a psychologist, there is intuitive plausibility to the idea that emotions are bodily feelings that are caused by various thoughts. This view helps us to make sense of the fact that emotions are related to both physical and mental events, and it is thereby particularly useful for relating the concepts of emotion and intelligence. According to it, emotions are no more or less rational, rash, immoral, or crazy than the thoughts that cause them. They need to be understood and evaluated. Often times they need to be acted on.

Ciarrochi et al. point out that it is exactly the repressive reactions that individuals have to their emotions that need to be overcome in striving for psychological health.[48] Salovey and Mayer similarly criticize the notion of an "emotional hijacking," as it merely replays the simplistic stereotype of emotions as an "intrinsically irrational and disruptive force." Anticognitive views of emotion implicitly, although perhaps unintentionally, encourage this type of repression.[49]

EMOTIONAL UNIVERSALISM AS EMOTIONAL INTELLIGENCE

A preliminary definition of emotional intelligence might sound something like this: emotional intelligence is the understanding of one's own emotions for the sake of acting on and expressing those emotions, seeking good outcomes and moral, psychological well-being, as well as understanding and discussing the emotions of others in the same pursuit and, additionally, creating an emotionally open and healthy environment that promotes emotional intelligence for all. This definition is based on the theory of emotional self-improvement that has been explored throughout this book. It contains all of the insights the theory of emotional universalism brings to the table: the need to understand emotions, the need to evaluate emotions, and the need to meet psychological needs. Furthermore, the score on which emotions are evaluated and oriented must be fundamentally moral.

Insofar as morality applies to human relationships, and humans, being humans, are emotional, it is morally necessary for us to cultivate emotional intelligence. We accept that the basic physical needs of food and shelter belong in a moral theory; the idea of emotional intelligence teaches us that people have psychological needs that are just as real and important as their physical needs. Emotional intelligence, then, must be provided for and expressed in any well-informed theory of the good life or right living. Being emotionally intelligent is related to being a good person in general. Being unaware of morality can lead to being emotionally cut off or to being emotionally off base.[50] In addition, not only does a lack of emotional intelligence cause bad behavior, but we need emotional intelligence in order to interact with people (including ourselves). We cannot promote the happiness of others or our own self-perfection without emotional intelligence.[51] As Aristotle notes, it is not even likely that we can comprehend the true content of our moral duties without something like a proper emotional comportment.

Psychological research into the skills necessary for emotional intelligence can help to complete a moral theory by providing instruction into the means of developing the ability to meet these moral and emotional needs. It is important to keep in mind just how simple an idea emotional intelliegence really is, and we must remember that our moral duties are embedded in the psychological reality of everyday life. As Zeidner, Matthews, and Roberts point out, there is some overlap between the idea of emotional intelligence and successful coping with stress.[52] Similarly, Brackett and Katulak's discuss the educative function of creating an emotional blueprint for a possible problem situation: "Creating an emotional blueprint" for a situation is just asking oneself questions about the emotions involved in a situation, such as "How may each person feel?" and "How may I feel?"; "What may the person be thinking as a result of these feelings" and "How might I respond to my feelings?;" "What may be causing these emotions?" and "How might these feelings be addressed?" These are ways of becoming more comfortable living with the challenges of human needs and interactions. (The father on the playground from the example in chapter 3 would have benefited from such an exercise.) In addition, we will also have to ask ourselves questions like "What is fair?" and "Am I respecting myself and others?" We must situate our discussion of emotional intelligence within the context of psychological and moral self-improvement. We might even find that emotional intelligence is improved by studying conflict-resolution strategies, stress-coping techniques, or by improving decision-making and critical-thinking skills.

Philosophically speaking, the real difficulty of emotional intelligence is figuring out why it is so *difficult* to keep in mind this simple, enhanced perspective about emotions and emotional responses. We could blame the emotions and say that emotions are inherently disruptive, but it is often times the case that people do not know what to do about emotions even when they are not having them. It is instead the case that emotions often stand in for deep needs and desires, and, although these needs may be simple, failing to meet them is a significant threat. Far from "hijacking" the "rational mind," perhaps emotions stop us from simply turning our back on problems we do not know how to solve.

Emotions often result from complex thoughts and problems. To fully understand them, or to be committed to trying to understand them, which is already a form of acceptance and expression, involves analyzing the way that they should be further expressed. Developing emotional understanding is difficult. If we think that emotions are always so easily understood and expressed, then we have misunderstood the nature of emotion and the meaning of expression. For example, suppose Pat hates his job—let's say he feels that the company is complicit in immoral deeds—but he is not financially able to quit and knows that his family and friends would not support his decision. He is likely to be sad and angry. Actually, to be sad and angry, he is relatively emotionally intelligent, since in this situation most people would be motivated by the unconscious desire to reduce cognitive dissonance and would deny or displace their sadness and anger, perhaps developing depression, an anxiety disorder, an addiction, or transferring these emotions to another person or situation. It is important that Pat remain confident in his understanding of his emotions, and that he not let external pressures make him lie to himself or ignore his emotional insights. Acting on the understanding of an emotion is not easy. (What *should* Pat do?) Struggling to do so, while resisting the urge to lie to oneself about the cause of the emotion in order to avoid addressing it, should be recognized as a genuinely difficult task.

Therefore, thus far we have seen that not only are we morally obligated to cultivate emotional intelligence—understanding, evaluating, and effectively acting on our emotions—but also we need to make use of moral evaluation in order to navigate our emotions and emotional situations. First and foremost we must have a moral orientation, being most concerned about what is the right thing to do, not an orientation toward happiness, pleasure, making money, finding fame, and the like; not even health offers a superior purpose in life. Although it might sound tautological that I should care first and foremost about what the right thing to do is (not, for example, what is best *for me*), having a moral orientation in one's life demands a fundamental transformation of self—a process that similarly requires emotional adeptness.

A study of emotional intelligence (if that term has not been twisted by Goleman to mean something like "leadership" or "business sense") is important philosophically. As many rationally inclined philosophers are quick to point out, intelligence is something for which we should strive and the notion of emotional intelligence suggests that simply having emotions is not good enough. We need to think more critically about emotional experience. Yet, if we insist on continuing to oppose reason and emotion, as Goleman does, we miss the significance of the intelligence of the emotions. The emotions are not "intuitions" from the "gut." Nor are the emotions intelligent in a way that externally supplements reason, as Greenspan, de Sousa, and Damasio, have argued.[53] Emotions should be seen as connected to many aspects of experience and inherent in processes of rational development. They result from not only latent, value-laden reasons for acting a particular way; they also come from deep-seated thoughts that refuse to be ignored. Some emotions speak for needs that might be denied, problems that must be solved, insights that must be recognized. Other emotions might stand in the way of doing what's right. To be emotionally intelligent means that we are adept in meeting the real moral challenges of life.

EMOTIONAL INTELLIGENCE AND MORALITY

Overall, in this book I have demonstrated a relationship between emotional experience and morality. Here we can see that the concept of emotional intelligence may involve little more than preexisting psychological topics (such as personality, family studies, and coping) with a normative spin;[54] in other words, this new trend in psychology is trying to accomplish the same thing that I have tried to accomplish here, although I am working from within a specifically Kantian framework. In this concluding section, I will list some of the ways that psychologists have already noted that the notion of emotional intelligence is a moral construct; then I will highlight some strengths Kantian moral theory adds to the theory of emotional intelligence. (I will not completely sum up my entire argument about the importance of Kantian moral theory; I will leave that to the fuller Conclusion in the next chapter.)

Many psychologists interested in the notion of emotional intelligence have already noted the need for moral theory to help define the construct. Waterhouse and Kristjánsson have both criticized Goleman's notion as potentially immoral.[55] The definition of emotional competence developed by Boyantzis and Sala, in conjunction with Goleman's model, which involves "an ability to recognize, understand, and use emotional information about oneself or others that leads to or causes effective or superior performance,"[56] lends itself just as easily to coercion, dishonesty, and greed as it would to honesty and fairness. Without a morally informed notion of emotional intelligence, nothing stops it from being the ability to effectively coerce and manipulate people. In her article on emotional intelligence in marriage, Fitness writes:

> It should be noted however that although this facet of emotional intelligence is potentially adaptive in marriage, someone who is skilled at reading other people's emotions could just as well use this ability for destructive as for constructive purposes. For example, married partners could conceivably use their empathic awareness in a calculated way to identify their partner's vulnerabilities and insecurities, and exploit these for their own purposes.[57]

Yet such guile should hardly count as emotional intelligence.

Saarni draws a connection between wisdom and emotional intelligence.[58] She defines emotional competence as "the demonstration of self-efficacy in emotion-eliciting social transactions." Self-efficacy means that the "individual has the capacity and skills to achieve the desired outcome."[59] Although it often goes unsaid, determining "the desired outcome," as with conflict resolution, comes under the purview of moral theory. Saarni further argues that moral character is a part of emotional competence, as she likens it to a virtue. She remarks that emotional skills, divorced from a moral sense, would not yield emotional competence because "emotional competence entails 'doing the right thing.'"[60]

Tugade and Fredrickson argue that emotional intelligence "is associated with higher quality interpersonal relationships among couples and friends."[61] Furthermore, positive emotions can produce increasing benefits over time and finding positive meaning amid stress can build personal resources, such as strengthened relationships and enhanced values (by inspiring moral courage, tolerance, and wisdom).[62] Those who work on emotional intelligence seem to agree that part of the goal of their work is to help improve people and help people improve their lives. Mayer writes:

> It is my hope that emotional knowledge will have a greater positive than negative impact. Societies that recognize the importance of their citizen's feelings may help create a more humane environment for those who live within them. When this emotional humanity is balanced with the other rights and responsibilities of the individual and society, the world may be better for it.[63]

Many components of emotional intelligence, like empathy, and its correlate self-respect, are moral notions. They are morally required, and we need other moral guides to tell us when we have achieved them. We do not measure empathy by the number of tears that we cry, but by making reference to ideals like equality and respect, as well as by considering our duties to promote the happiness of others and our own self-perfection.

James Averill likens emotional intelligence to virtue by drawing on Aristotle:

> About 2300 years ago Aristotle observed that emotions "may be felt both too much and too little, and in both cases not well; but to feel them at the right times, with reference to the right objects, towards

> the right people, with the right motive, and in the right way, is what is both intermediate and best, and this is a characteristic of virtue" (*Nicomachean Ethics* 1006b20). More than of virtue, this is a good description of *emotional intelligence.*

Nevertheless, Kant's notion of virtue is more conducive to developing emotional intelligence than Aristotle's. Aristotle himself goes beyond describing the virtuous amount of feeling as the moderate amount of feeling; he, like Kant, also tells us that the virtuous person *understands* what the right amount of feeling is and why. Aristotle does not share that theoretical understanding of the virtuous person with his reader; he believes that it is something that is impossible to teach intellectually or abstractly. The best we can do is to find a virtuous person to model. Nevertheless, it does seem that we can gain intellectual insight into the best way to express our emotions, and Kant gives us a theory that can better help us evaluate our behaviors.

For starters, Kant encourages us, in everything that we do, to treat people—including ourselves—with respect. The right emotions will line up with this moral requirement. Kant also argues that we need to think about the consequences of our actions and universalizability of our principles. We cannot divorce one instance of an action from the social context in which it will be encoded and repeated. Even within our own thinking, we rely on principles. We often make choices that are easy for us in the short term, but we try to ignore the principle we are thereby affirming and the possible negative consequences to others and perhaps even ourselves. Our emotions then tend to defend these shortsighted whims, building a shallow, repressive character overall. Instead we need to develop moral emotions, like conscientiousness, courage, cheerfulness, and patience, that support a commitment to living virtuously. We must seek to understand our own motivations, making sure we rise above selfishness, and we must seek to promote the happiness of others.

Aristotle agrees with this higher calling of life; happiness for him is more than money, fame, or pleasure. All of these goals, while they have a place in life, leave one with the sinking feeling that life is meaningless. Moral goodness, on the other hand, reveals the meaning of life. Accordingly, as we have seen, we need moral judgment to help us hammer out the very idea of emotional intelligence. Kantian moral evaluation can help us hammer out the moral questions that accompany everyday emotional experience, questions like: Should I favor confrontation at any cost? Or emotional conformity? Should I express this emotion although it will hurt another person? Should I promote this relationship or end it? Should people lie to themselves in order to protect their self-esteem and established beliefs or to reduce cognitive dissonance? Or should honesty be the highest virtue (as it appears to be for anxiously attached couples[64])? Emotionally intelligent behavior involves the evaluation of emotions: if emotions are not evaluated by moral standards, then any standards on which they are evaluated will lack an ultimate foundation.

Kantian moral theory steers us toward recognizing inherent human dignity—hence the need to respect people (including ourselves) and treat them fairly—as well as the connection between moral achievement and self-esteem. It counsels us to stick up for ourselves as well as reach out to others. It assumes that we are often selfishly trying to get away with something and cautions self-criticism, but it also counsels that if we do not feel the love, pain, respect (and many other feelings) associated with morality, we risk becoming morally dead.

An emotional intelligence program that is not fused with moral theory often becomes too centered on the subjectivity of emotions and fails to address the real problems that the specific emotional situations present. Also, it appears dumbfounded in the face of the moral crises at the heart of many emotional experiences: calls like "Am I being treated fairly?" or "Is it okay for me not to sacrifice what I want?" In other words, a notion of emotional intelligence that ignores moral questions risks being ineffective (if not harmful). Whereas a therapist who is adept in moral theory can simply jump into the dialogue in the heart of the emotion.

Conclusion

The main arguments of this book can be summed up by answering the questions "What about Kant's philosophy is helpful when discussing emotion?" and "Specifically, how does it help?"

I have drawn from both Kant's philosophy of emotion and moral theory to show that they are helpful in evaluating emotion.

Kant believes that emotions are physiological events, however slight, that follow from certain mental events. Perhaps we should say that the emotion is the conjunction of the thoughts with its bodily effects because the physical events make little sense without an understanding of their mental causes and not all physiological feelings are emotions, only those that are caused by mental events are.

This theory of emotion is helpful for a number of reasons. The main reason is that it creates a space for a great diversity of emotions and explains the need for emotional understanding and evaluation. Because emotions are related to different types of mental events, it is impossible to judge them wholesale—to say that they are all troublesome or all useful, for example. Just as not every thought we have is accurate, virtuous, natural, logical, or helpful, not all of our feelings are either—but some are. The only way to know whether or not an emotion represents any of these good qualities is to look at the meaning behind it. Simply feeling good or not feeling good is insufficient.

Kant's robust account of moral feelings is also helpful for setting us on the path to develop virtue. It teaches us that we already have a number of moral feelings, and we need to open ourselves up to having more of them. As with realizing that emotions are no better or worse than our thoughts, the insight that we have many moral feelings sets us up to accept emotions in ourselves and others. Human beings have many natural needs and being sensitive to and respecting these needs puts us in the realm of feeling many things—fear for my well-being, love for my children, passion for learning, anger at injustice, sympathy and sorrow at misfortune, among many examples. Accepting these emotions is a necessary part of working to make sure we live in a world in which care is taken for everyone's needs.

Kant's theory of emotion is often misunderstood because it is overshadowed by his constant injunction to morally evaluate our desires. Emotions are not the same thing as desires, but emotions can be a part of creating desires, making choices, and enacting them. Kant's moral theory highlights the corrosive pull of selfishness and false goods; universalism and moral commitment are the antidote. While we must be on guard against selfishness and imperfectly moral pursuits, there is little reason to conclude thereby that Kant thinks that all emotions are selfish or superficial. In fact, more often than not, denying or ignoring one's emotions or the emotions of others involves denying moral duty.

Kant's notion of universalism is helpful because it highlights the role that maxims, or principle, plays in our lives. Acting in a way that we cannot affirm as a principle introduces cognitive, emotional dissonance, that is, internal conflict. Throughout this book I have suggested ways in which we have moral feelings, yet we often suppress them. Expressing and acting on moral feeling takes courage; nevertheless the impetus to test our principles for universalizability as well as universal respect is already present as a "fact of reason."

Kant's notion of moral commitment and self-esteem is useful because it ties value to moral value, giving us a clear sense of what we need to do in order to feel good about ourselves. It is no doubt the case that many people generally resist moral commitment. Nevertheless, I am hopeful that as people become more educated about the consequences of their choices and the general state of the world, they will not resist transitioning to an objective standpoint of moral responsibility.

The fringe idea that people should be selfish has become more mainstream recently. Extreme neoliberalism has borrowed from Ayn Rand the acceptability of selfishness. This has also come to pass because some social scientific theories of human nature, like rational actor theory from political science and economics, has been misconstrued, in the absence of any other knowledge of moral theory, as itself a moral theory. When I was in college, I believed that there was no such thing as selfishness. I am not quite sure why. I think it was because I was held by the sway of notions of enlightened self-interest, which hold that anything purely selfish is not really good for you anyway. Only kind and generous things are actually good for you, and, hence, selfish. I no longer think that. Now, having been convinced by Kant, the position that doing what is good is actually *good for you* anyway (in some sense or in the long run) seems to be morally corrupting. The same goes for the notion of karma. Although it is quite difficult for normal people to accept, since it violates our hopes for religion and cosmic justice, bad things happen to good people. Similarly, morality often calls for the sacrifice of some things that give us pleasure. It is true that virtuous people might experience a *qualitatively* greater pleasure in achieving noble goals, or even just in attempting them, but we cannot overlook the extent of the real pain and sacrifice that is sometimes required.

Kant's understanding of emotion not only helps to correct the simplistic and unhealthy approaches to emotionality that were canvassed in the first chapter, but it also helps to criticize another current trend in understanding emotion, namely, neurobiological reductionism. We saw some cases of this trend in chapter 3. Neurobiological accounts have the capacity to dazzle us into thinking that something more is being demonstrated than simply the site in the brain that corresponds to certain behaviors or mental events. Similarly, since the brain itself is taken to be a cause, neurobiological therapies can thereby appear attractive. Nevertheless, as Nussbaum has argued, there is little that brain research can do to contradict mental experience.

While Kant's theory of emotion is helpful in this regard because it demonstrates the connection between cognition and physical feeling, some of Aristotle's remarks are also necessary for keeping a clear head when considering the brain. Aristotle's four causes—material, formal, final, and efficient[1]—help us to understand that there are, what we might call, different levels of reality and, as convincing philosophers of mind, like Spinoza, James, or Davidson have demonstrated, we need not hold either the material or mental level to be more primary. Brain scientists, like other scientists, explore material and efficient causality. In other words, they show us the material states that cause certain events and the material states that materially caused those material states. Sometimes the material and efficient causes are the types of causes we're asking after. If I ask, "Why is a ruby red?" for example, I most likely want to know about its material/chemical composition. If I ask "How is vision possible?" I also most likely want to know about material and physical processes, this time in the brain. Nevertheless, if I ask "How is morality possible?" or "What is emotion?" I may or may not be looking for a material answer, depending on my purposes, and a material answer does not serve all purposes. If I want to make sure everyone in my community can develop moral goodness, I need a complete account of morality and emotion: I need to provide not just for the physical component of morality and emotion, that is, proper nutrition, and so on, but I need an account in terms of formal and final causality—a philosophical account like the one offered in this book.

Aristotle also argues that different types of scholars study phenomenon in different ways:

> Accordingly, a physicist and a dialectician would define each [attribute of the soul] in a different way. For instance, in stating what anger is, the dialectician would say that it is a desire to retaliate by causing pain, or something of this sort, whereas a physicist would say that it is the rise in temperature of the blood or heat round the heart. (*De Anima* 403a25–33)

Neurobiologists focus on the brain and philosophers focus on rational comprehension. It is not a problem—in fact it is good—that each does a different job. While in some rare cases one might cause problems for the other—if for example there were never any neocortical activity during any emotional experience[2]—it is safer for each just to do her own job in her own way and not pretend to be the expert on everything. There is danger in reductionism, either biological or philosophical, as we thereby lose an appreciation of the different types of study. Furthermore, if we consider ourselves only from the material perspective and not as rational beings, we are offending our moral self-worth.

Starting from a scientific approach can confuse us about the nature of morality. Prinz uses an equation between metaethics and moral psychology, that is, his theory of emotion, as a way to bridge the Is-Ought boundary.[3] His goal is to reconcile naturalism and normativity. He believes that we are morally obligated to figure out what we personally believe is good so that we can better follow it. Prinz's assumption that "if naturalism is right, then moral facts are natural facts, or they are not facts at all" seems reasonable on first blush. He then concludes that if we want to hold onto morality we need to break Hume's Law. It seems right to me that all of our observations about the way that the world IS, that is, our metaphysical claims, should be explained naturalistically. It then seems to follow that the concept "ought" must be analyzed naturalistically, and the only way to do that is in terms of psychology, translating it into "prescriptive sentiments." This is the way we explain the cause of natural objects—we give the material and efficient causes. Nevertheless, moral theory need not be metaphysics. We need not have recourse to the idea that there IS such a thing as The Good, that is, that it *exists*, in order to have a concept of The Good. As Kant recognizes, the important matter is not the generative account of these concepts (*quid facti*), but the justification for these concepts (*quid juri*). *That* we have them (and where we got them from—our parents, religion, reading Kant, etc.) is a trivial point; what matters is that they make sense, that is, that they are right. We do not judge the veracity of moral concepts by means of a correspondence theory of truth; doing so would require metaphysics. We rationally evaluate them.

If we fully accept that The Good is not real in the same sense that a physical object is real, then we can sidestep Rawls's argument about the different between moral intuitionism and moral constructivism. Rawls argues that because Kant does not hold that The Good really exists as an external object that we must intuit then The Good must be "constructed" out of reason itself.[4] The soundness of this argument hangs on the meaning of the term "construction." If it denies rational objectivity to the concept of the Good along with physical objectivity or if it conflates moral and prudential reasoning, then the argument is invalid and based on a false dichotomy. (Being objectively real is related to being constructed out of moral [or universal] reasoning, not just subjective, pragmatic reasoning.)

Do we "intuit" that 2 + 2 = 4? Do we look up in the sky and see it written there? No, rational entities are not physically external entities, but that does not mean that they do not exist. I once had a student who argued that because people are free and can choose to do whatever they want to, morality does not exist. In other words, since the world IS not moral, then there is no OUGHT. This argument is similar to Prinz's thinking. The problem with it—aside from the fact that it fundamentally misunderstands the nature of morality—is that there are lots of oughts. Humans are moral animals: you ought to smile; you ought to go to school; you ought not have sex before marriage; you ought to floss your teeth; you ought to be passionate and enjoy life; you ought to drive at the speed limit; you ought to dress fashionably; or, Prinz's injunction, you ought to be consistent. We are bombarded with oughts, and we cannot live without some means of evaluating them. We evaluate them in terms of a personal hierarchy of values, but how do we evaluate that personal hierarchy? How do I know if cleanliness, for example, really is the kind of thing that I should value? At that point, I need a moral theory: I can look at principles, consequences, ideal worlds, biblical stories, or some combination of these, but without some account of what the Good is, I cannot make sense out of the worth of other goods. Of course, it is complicated, but we make sense out of it the same way we humans make sense out of everything, by thinking about it.

As I have shown, the question of what we should do is especially relevant to the experience of emotion. Emotional universalism offers us a theory about the way we should evaluate our emotions. The three senses of emotional universalism parallel three steps of emotional evaluation: accepting and understanding our emotions, canvassing our emotions *about* our emotions while committing to integrate these insights into our moral lives, and recognizing that our moral lives must be emotional. Furthermore, we can put what we have learned from Kantian moral theory into simple, emotionally relevant rules: (1) Be honest with yourself (strive as much as you can for self-knowledge and self-scrutiny); (2) put morality first (true self-esteem looks toward goodness, not toward happiness); (3) accept the equal value of others (you are no more or less important than anyone else, you are just one human being). These three take-home points relate to major themes in Kant's moral theory, and, additionally, I have shown how the content of Kantian moral theory, namely, the categorical imperative, can be used to further evaluate our emotions. We must examine whether or not our emotions harbor selfishness and self-exception. We must examine whether or not an emotion respects ourselves and others. Furthermore, we must seek to promote our own self-perfection and the happiness of others.

All of this advice might still seem overly abstract when we are in the throes of an emotional crisis. In those cases, perhaps it will help to call to mind the basic message of emotional universalism: emotions are normal, everyone has them; nevertheless, we must work to integrate them into our lives. We are tempted to ignore them until they go away, but ignoring emotions means ignoring the most important aspect of our lives, our interpersonal and moral connections.

Notes

Introduction

1. I am grateful to Frederick Beiser for his helpful comments on this chapter.
2. Although feelings are of two kinds, according to Kant (sensuous and intellectual) sensuous feelings include feelings that arise from the five senses and those that arise from present experience (what he calls the faculty of "imagination" because it creates [real] images). For example, present experience can make one feel bored or amused (A 233). My assumption is that feelings arising from present experience count as "emotions" for us. Kant makes a distinction between what he calls the imagination/understanding and the intellect/reason, so describing Kant's theory of emotion as a theory of "intellectually wrought feelings" is not as accurate as locating "emotion" within his theory of "cognitively" or "mentally" wrought feelings.
3. It is true that many folk theories of emotion as well as scientific theories take the two to be integrally related. Nevertheless, we should note several reasons in favor of such a distinction: (1) Kant classes feeling and desire as separate faculties. (2) It may be necessary not to act on (repress) certain desires, but suppressing or repressing certain thoughts will not help in facilitating learning or moral evaluation. (More evidence against thought suppression/repression comes in chapter 1.) (3) As Sorensen notes, although feelings often lead to desires, they need not, and Kant highlights a number of feelings, like aesthetic feeling, that are disconnected from desires. See Kelly Sorensen, "Kant's Taxonomy of Emotions," *Kantian Review* 6 (2002), 109–128. Nevertheless, pleasure and pain might seem to be naturally connected to desire; recognizing this might help us to explain the difficulty that occasions the experience of negative emotions. Even still, although odd in theory, it is perhaps common to have difficulty experiencing positive emotions, which suggests that the thought content behind intellectually wrought feelings can be philosophically complex.
4. Geiger does a good job of mapping this distinction for Kant. See Ido Geiger, "Rational Feeling and Moral Agency," *Kantian Review* 16 (2011): 283–308.

5. Herman's rich notion of Kantian judgment can be developed to the point of illustrating the lived connection between reason and emotion, as well as their natural moral development, but she nevertheless still overlooks Kant's theory of feeling in favor of his theories of desire and judgment. See Barbara Herman, "Making Room for Character," in *Aristotle, Kant, and the Stoics: Rethinking Happiness and Duty*, ed. Stephen Engstrom and Jennifer Whiting (New York: Cambridge University Press, 1996).
6. Much attention has been given to the difference between moral motivation and amoral motivation and the role that desire (thought to be synonymous with emotion) is allowed to play in moral motivation. This is understandable because Kant expresses much concern about the inclination to violate moral duty and so commentators rightly focus on desire when considering Kantian moral deliberation. Nevertheless, this focus misses the most sure means of cultivating virtue, namely, attention to feeling. Kant is right that immoral desires are part of the problem, but focusing on moral feeling is a surer solution.
7. With Kant's reference to respect in the *Groundwork for the Metaphysics of Morals*, he offers a qualification: "But even though respect is a feeling, it is not one received through any outside influence but is, rather, one that is self-produced by means of a rational concept; hence it is specifically different from all feelings of the first kind (*der ersteren Art*), which can all be reduced to inclination and fear" (G 401). This qualification might confuse the reader into thinking that Kant is making a distinction between respect and all other feelings. Actually, in *Anthropology from a Pragmatic Point of View*, he make a distinction between two types of feelings: (1) those based on sensuous pleasure, and (2) those based on intellectual pleasure. Intellectually wrought feelings are divided into those that come from concepts and those that come from ideas. Kant's reference in the *Groundwork* to feelings "of the first kind" most likely refers to the distinction made explicit in the *Anthropology*. Hence, while respect is undoubtedly the most important feeling for Kant, there is no reason to think that respect is the lone instance of an intellectually wrought feeling. (It is unfortunate that Kant in the previous passage appears to class inclination and fear as sensuous feelings, since fears can arise from concepts as well as ideas—the feeling of respect even involves a degree of fear. An inclination, on the other hand, is defined as a base impulse, as opposed to an "interest" (G 413). Perhaps, Kant would have been better off to choose the word "aversion" here, instead of "fear," since we have a natural inclination toward physical pleasure and a natural aversion to physical pain.)
8. Kelly Sorensen, "Kant's Taxonomy of Emotions."
9. Although the terms "ethics" and "moral theory" are often used interchangeably in English, for Kant, "morality" is the larger term that also implies a consideration of the theory of justice, that is, laws that can be fairly externally coerced regardless of the individual's mindset, along with the theory of virtue. Ethics refers to "internal lawgiving" (DR 220). Hence, in the *Metaphysics of Morals* (*Metaphysik der Sitten*), Kant covers

both those actions that can be publicly coerced and those maxims that are private. (For evidence that the "supreme principle of morality (*Moralität*)" grounds the whole of the *Metaphysics of Morals*—the "Doctrine of Right" as well as the "Doctrine of Virtue"—see G 392.) On the contrary, ethics for Kant is a consideration of what individuals should take up as their goals (including obedience to just laws). Although it can apply to anyone, it assumes a first-person perspective, or what Kant calls "internal law-giving" (see DV 218–221). Aristotelian virtue ethics also takes up the same first-person perspective, but utilitarianism comes at questions of behavior from a more objective, third-person perspective, and like Kant's moral theory overall, it includes questions of justice and external coercion.

10. "A good will is good not because of what it effects or accomplishes, nor because of its fitness to attain some proposed end; it is good only through its willing, i.e., it is good in itself" (G 394).
11. Kant identifies "a pure will" with one that is determined, not empirically (by a concern for particular effects), but by a priori moral principles (G 390).
12. Finding the principle that best expresses the action is difficult. Some children at my house were throwing toys around. I asked them, "Would you do that if they were your toys?" I actually did not think that it would cause any problems to throw the toys, but I suspected that *in their minds* they were showing disrespect and trying to get away with something. Another example of the difficulty in seeing the principle behind a situation: I left my sled at my son's elementary school during the day although clearly there is not enough room at school for *everyone* to leave their sleds there. One might object that I was taking advantage of the situation and trying to get away with something. That is not really true: everyone could *in principle* leave their sleds at the school; there is just not enough room, and in the event that everyone wanted to leave their sleds at the school, we could find a practical solution to that merely practical problem, like making a sled parking lot, erecting a shed, or alternating days on which people were allowed to bring sleds. To make a parallel with the toy throwing, I might ask myself, "Would I think it is okay if someone else left their sled outside the school?" and the answer is "Yes."
13. Kantian moral theory does not assign inherent value to nonhuman animals, only instrumental value, but it might not be difficult, based on my explanation here (of the connection between the necessity of inherent goodness for moral considerations and universal inclusion) to widen the sphere of moral consideration to animals. Hedonistic utilitarianism, which has an easy time including animals, holds that all pain is morally bad, wherever it occurs, but Kantian moral theory does not hold that pain is bad in itself. Indeed, undergoing pain might sometimes be morally necessary. Although I think (along with Kant) that it is morally wrong to cause unnecessary suffering to animals or to fail to provide for their well-being, when possible, I am agnostic about whether or not there is

any problem with a human-centered moral theory such as Kant's, especially because Kant's theory can be seen to include these animal moral considerations in other ways (see DV 443). I do think Kant is right to believe that morality is exclusively human as it is related to the rational sphere of governance. That exclusivity helps to explain the paradoxical fact that there is nothing immoral about animals killing other animals, however distasteful it is to us; although there is most assuredly something morally wrong with humans killing other humans, despite the fact that we are also animals, and there might even be something morally wrong with humans killing animals, depending on the way it is done.

14. I agree with Denis that the notion of duties to oneself, as it involves the development of moral commitment, is foundational in Kant's ethics; Lara Denis, *Moral Self-Regard: Duties to Oneself in Kant's Moral Theory* (New York: Routledge, 2001).
15. We sometimes consider laws of justice as abstracted from the perspective of virtue, but our comprehension of morality underlies both justice and virtue: "Whereas there are many duties of virtue... there is only one virtuous disposition, the subjective determining ground to fulfill one's duty, which extends to duties of right as well although they cannot, because of this, be called duties of virtue" (DV 410).
16. Henry Allison, *Kant's Theory of Freedom* (New York: Cambridge University Press, 1990), 130.
17. I discuss whether or not there are nonselfish types of immorality in chapter 6. Denis outlines Kant's distinction between "selfishness" and "rational self-love," writing, "selfishness leads one to think that what one desires on the basis of inclination is objectively good." Denis, *Moral Self-Regard*, 165.
18. Kant speaks against the "exaggerated discipline of one's natural inclinations" and states instead that we should provide for our own "cheerful enjoyment of life" (DV 451).
19. Kant sides with Epicurus's notion of "the ever cheerful heart" of virtue over the stoic endurance for suffering in his section on "Ethical Ascetics," where the term ascetics only means practice and self-training (DV 484–485).
20. See Johnson's discussion of Kant's philosophy of nonmoral self-improvement; Robert Johnson, *Self-Improvement* (New York: Oxford University Press, 2011).
21. "Adversity, pain, and want are great temptations to violate one's duty. It might therefore seem that prosperity, strength, health, and well-being in general, which check the influences of these, could also be considered ends that are duties, so that one has a duty to promote *one's own* happiness and not just the happiness of others. But then the end is not the subject's happiness but his morality, and happiness is merely a means to removing obstacles to his morality—a permitted means, since no one has a right to require of me that I sacrifice my ends if these are not immoral. To seek prosperity for its own sake is not directly a duty, but indirectly

it can well be a duty, that of warding off poverty insofar as this is a great temptation to vice. But then it is not my happiness but the preservation of my moral integrity that is my end and also my duty" (DV 388).

22. I use the term "moral feeling" to refer to both feelings that arise from moral actions and feelings that arise from moral thoughts alone.
23. "For a human being cannot see into the depths of his own heart so as to be quite certain, in every *single* action, of the purity of his moral intention and the sincerity of his disposition" (DV 392).
24. It is my opinion that human ineffectiveness, the examples of which are painfully multitudinous, is more often the result of lack of unselfish vision and resolve—which are themselves the result of a lack of self-understanding and moral commitment—than of a lack of intelligence.
25. Carol Hay defends the method of "abstraction" employed in universalism from the criticism that it abstracts from morally relevant information, like emotion. See Carol Hay, *Kantianism, Liberalism, and Feminism: Resisting Oppression* (New York: Palgrave Macmillan, 2013), 20. In addition, Hay as well as Peter Singer do a good job of providing an impersonal justification for certain kinds of preferential treatment. See Hay, *Kantianism, Liberalism, and Feminism*, 59, and Peter Singer, *One World: The Ethics of Globalization* (Princeton, NJ: Princeton University Press, 2002), 160–167.
26. Herman's interpretation of practical reason and judgment is helpful here. See Barbara Herman, *The Practice of Moral Judgment* (Cambridge, MA: Harvard University Press, 2007).
27. I take for granted that, as Herman put it, "the requirement of salience—much of the work of moral judgment takes place prior to any possible application of rules." In other words, following McDowell, she sees that "the complexities of moral judgment involve the complexities of a developed moral character." Herman, "Making Room for Character," 36.

1 PROFILES OF EMOTIONALITY

1. Miriam Griffin in "Cynicism," *In Our Time*, BBC Radio 4, October 20, 2005.
2. *Seneca's Epistles*, cxiii, Sec. 28. Also, "There is nothing grand that is not also calm" (Seneca's Dialogues, Book iii, Chap. Xxi, Sec. 4); See Frederic Holland, *The Reign of the Stoics* (New York: C. P. Somerby, 1879).
3. Nancy Sherman, *Stoic Warriors* (New York: Oxford University Press, 2005).
4. Epictetus, *Handbook of Epictetus*, trans. Nicholas P. White (Indianapolis: Hackett, 1983), III.
5. See Andrew M. Holowchak, *The Stoics: A Guide for the Perplexed* (New York: Continuum, 2008). Sandbach argues that the Stoics did not uniformly reject emotion because they do hold that there are good emotions (*eupatheiai*). Instead, privileging the theories of Greek over Roman

Stoics, he argues that they only criticized "passions," which are by definition irrational. See F. H. Sandbach, *The Stoics* (Indianapolis: Hackett, 1994). In any case, it is undeniable that the Stoics held that virtue is entirely within one's control and, therefore, there is no such thing as physical goods. If emotions speak for physical and external psychological needs, then the Stoic will necessarily classify them as irrational. Nevertheless, my goal here is not to unassailably interpret Stoic philosophy but to illustrate a common emotional comportment, which we can call "Stoic" easily enough.

6. See Daniel Goleman, *Emotional Intelligence: Why It Can Matter More Than IQ* (New York: Bantam Books, 1995).
7. See Sherman, *Stoic Warriors.*
8. Psychologists often call violence caused by suppressed anger "eruptive violence." Of course, without a good understanding of conflict resolution, anger can also lead directly to violence. Nevertheless, there is no necessary connection between anger and violence. Robert F. Marcus, "Emotion and Violence in Adolescence," in *Encyclopedia of Violence, Peace and Conflict*, ed. Lester R. Kurtz (Waltham, MA: Academic Press, 2008).
9. For a good review of the literature, see Christine Purdon, "Thought Suppression and Psychopathology," *Behavior Research and Therapy*, 37 (11) (November 1999): 1029–1054.
10. For a good account of Kant's dismissal of the Stoic understanding of happiness, see Frederick Beiser, "Moral Faith and the Highest Good," in *The Cambridge Companion to Kant and Modern Philosophy*, ed. Paul Guyer (New York: Cambridge University Press, 2006).
11. David Schnarch, *Passionate Marriage* (New York: W. W. Norton, 2009).
12. Jürgen Habermas, *The Philosophical Discourse on Modernity* (Cambridge: MIT Press, 2000).
13. L. Aknin, M. Norton, and E. Dunn, "From Wealth to Well-Being? Money Matters, But Less Than People Think," *Journal of Positive Psychology*, 4 (6) (2009): 523–527.
14. She reviews the scientific evidence against the popular misconception that "positive thinking" can prevent and treat diseases, like cancer. Barbara Ehrenreich, *Bright-Sided* (New York: Metropolitan Books, 2009). I am grateful to Kelly Sorensen for this reference. In fact, the core assumption that the mind can control the body, willing a stronger immune system, for example, is reminiscent of Christianity and Stoicism, whereas research into the physical causes of diseases has routinely been successful.
15. One example: Erica Schütz et al., "The Affective Profiles in the U.S.A.: Happiness, Depression, Life-Satisfaction, and Happiness-Increasing Strategies," *PeerJ*, available at: https://peerj.com/articles/156/. This study, like others, assumes that values are subjective rather than potentially objective and universal.

16. As the show progresses, Chris begins seeing a therapist and learns how to get in touch with his negative emotions.

2 Understanding the Nature of Emotion

1. I am grateful to Ray Boisvert for his helpful comments regarding this chapter and for giving me the idea for chapter 1.
2. A close look at the *Philosophical Investigations* reveals that Wittgenstein intended to imply that all concepts are family resemblance concepts, although some "families" might be more inbred than others. His point is that we must get at meaning by looking at use; multiple utterances of a term are related to each other through the recognition of the similarity of context. Given that meanings change overtime, we cannot draw a hard-and-fast line between a metaphorical and a literal use, an utterance can only be more or less literal or metaphorical given the similarity of the contexts, but no two contexts are identical.
3. Solomon and Calhoun write that, for James, an emotion is a "physiological reaction." See "Introduction," in *What Is an Emotion?* ed. Robert Solomon and Cheshire Calhoun (New York: Oxford University Press, 1984), 3. James's essay is reprinted in this collection, and the page numbers I give refer there.
4. I will not follow this terminology. Instead, I tend to use the term "feeling" to refer to a physical sensation and "emotion" to refer to the whole complex of situation, thoughts, and feelings. Nevertheless, I do not find any specific problem in using the two terms as synonyms, as we normally do in everyday speech.
5. Leda Cosmides and John Toobey, "Evolutionary Psychology: A Primer," available at: http://www.psych.ucsb.edu/research/cep/primer.html
6. Joseph LeDoux, *The Emotional Brain: The Mysterious Underpinnings of Emotional Life* (New York: Simon & Schuster, 1998).
7. William James, "What Is an Emotion," in *What Is an Emotion?* ed. Cheshire Calhoun and Robert Solomon (New York: Oxford University Press, 1984). Originally published in *Mind* (1884).
8. That we find certain baby-like things cute, ranging from baby animals and pets to stuffed animals and pet rocks, might count as evidence of an innate and easily misplaced affection for human infants. Nevertheless, the lack of universality of the cute response, as well as great variation in its expression, would count as counterevidence.
9. Richard Lazarus offers an appraisal theory of emotion, much like the cognitive theories we will discuss shortly although with a biological focus. See Richard Lazarus, *Emotion and Adaptation* (New York: Oxford University Press, 1991).
10. Paul Ekman's research on facial expressions supports the conclusion that there are universal emotional responses, or core relational themes, but, for him, the triggers of those responses are culturally conditioned.

Sadness may be universally experienced after a personal loss, for example, but the event that counts as a personal loss, will be, to some degree, subjective and culturally relative. See Paul Ekman, *Emotion in the Human Face: Guidelines for Research and an Integration of Findings* (Oxford: Pergamon Press, 1972).

11. Prinz also makes use of Damasio's idea of the as-if loop to explain instances where emotions "bypass the body." See Jesse Prinz, *Gut Reaction: A Perceptual Theory of Emotion* (New York: Oxford University Press, 2004), 72. It is not clear whether this phrase is meant to refer to emotions that occur without affects or to all emotions that are triggered by judgments.
12. Cannon was the first to express this argument that "the same visceral changes appear in a number of different emotions" (18). W. B. Cannon, "The James-Lange Theory of Emotion: A Critical Examination and an Alternative Theory," in *The Nature of Emotion*, ed. M. B. Arnold (New York: Penguin, 1968).
13. S. Schachter and J. E. Singer, "Cognitive, Social, and Physiological Determinants of Emotional State," *Psychological Review*, 69 (5) (1962): 379–399. For my purposes here, the Singer-Schachter "two factor theory" still counts as an affective theory of emotion because it tends to assume that emotions first and foremost are a response to external stimuli. They begin to cross the divide into cognitive theories when they acknowledge that interpretations of the arousal can themselves cause arousal, but in so far as they would see those interpretations as secondary, they are still in the affective camp.
14. Prinz, *Gut Reactions*, 30–31.
15. There seems to be some confusion here about whether or not neurological responses count as physiological responses. James focused on symptoms in the internal organs, but later affective theorists focus on the brain. Prinz seems to want to have it both ways, to refer to "gut reactions" and also neurological perceptual mechanisms. In other words, there seems to be some confusion between the idea that emotions are primarily reactive affects and the privileging of an examination of the material cause of emotion (in the brain). No one would deny that emotions have some kind of neurological correlate, but neurobiologists cannot use the idea that emotions primarily are physiological responses in order to defend the priority of neurobiological research and vice versa.
16. For an argument that Aristotle's theory does represent purely cognitive theory and that it is comprehensively constructed, see Chapter 1 of W. W. Fortenbaugh, *Aristotle on Emotion* (New York: Barnes & Noble Books, 1975).
17. The context in which we find the lion's share of Aristotle discussion of emotion (pathos) is particularly significant and is mostly responsible for the fact that Aristotle is thought to have a cognitive theory of emotion. The goal of the *Rhetoric* is to teach lawyers how to sway jurors. It is no surprise that an "Aristotelian" theory of emotion holds that emotions are

beliefs that are based on some kind of evidence: the giving of evidence is the only means a lawyer has of stirring emotions. Furthermore, jurors are in a position to judge about a defendant, although not to actually interact with the defendant. Similarly, Aristotle's discussion of emotion portrays emotions as *dispositions* to actions, such as the disposition to help another with no benefit to oneself (*charis*), that are disconnected from the actions themselves: one can feel *charis* (or gratitude) toward a person without doing anything about it. (See David Konstan, "The Emotion Is Aristotle Rhetoric 2.7," in *Influences on Peripatetic Rhetoric*, ed. David Mirhady [Amsterdam: Koninklijke Brill NV, 2007], for an argument that *charis* should be translated as gratitude.) It is possible that the problem in translating this term comes from exactly the point about context to which I refer, namely the fact that charis is an emotion that is normally connected to some action (as is grace, or *gratis*) but is disconnected from its action in this context. The juror is in a position to act regarding the defendant but in a way that is removed from a direct relationship.

18. John M. Cooper, *Reason and Emotion: Essays on Ancient Moral Psychology and Ethical Theory* (Princeton, NJ: Princeton University Press, 1999), 422. Note that this three-part description is similar to Dewey's but Dewey focuses on the consciousness of the feeling, not the consciousness of the situation that leads to the feeling. See John Dewey, "The Theory of Emotion. (Part II) The Significance of Emotions," *Psychological Review*, 2 (1895).
19. Aristotle also considers one way that emotions relate to desires: anger is usually the result of a frustrated desire, and so one is "carried along by his own anger by the emotion [the desire] he is already feeling" (Rhetoric 1379a13).
20. Cooper argues that Aristotle's long list of emotions in the *Rhetoric* is merely a "dialectic investigation" meant to prepare the way for a "scientific" theory, and not such a theory itself. Cooper, *Reason and Emotion*, Chapter 19.
21. Sokolon's argument that Aristotle believes that they can also sometimes be rational is obviated by a clarification between the irrational and nonrational. Something that does not originate from reason (the nonrational) need not be opposed to reason (irrational). See Marlene Sokolon, *Political Emotion: Aristotle and the Symphony of Reason and Emotion* (DeKalb: Northern Illinois University Press, 2006), 19.
22. *Pathos* is better translated as "passion" than as "emotion," but this does not stop contemporary philosophers of emotion from borrowing from either Aristotle or the Stoics. See F. H. Sandbach, *The Stoics*, 2nd ed. (Indianapolis: Hackett, 1989), 59–68.
23. Sherman argues that Buddhism also holds that reason can itself be a dangerous object of attachment. See Nancy Sherman, *Making a Necessity of Virtue: Aristotle and Kant on Virtue* (New York: Cambridge University Press, 1997), 115. Therefore, Buddhism is more like Pyrrhonism, which

also aims at ataraxia, than Stoicism. Still, the Pyrrhonist slogan *ou mallon* (more or less) is similar to the Stoic ideal of apathy.

24. Martha Nussbaum, *Upheavals of Thought: The Intelligence of Emotions* (New York: Cambridge University Press, 2001), 64.
25. Robert Solomon, "Emotions and Choice," in *Explaining Emotions*, ed. Amélie Oksenberg Rorty (Oakland: University of California Press, 1980), 257.
26. Robert Solomon, "Emotions, Thoughts, and Feelings," in *Thinking About Feeling*, ed. Robert Solomon (New York: Oxford University Press, 2004), 77.
27. This is Sartre's Heideggerian language, adopted by Sartre to respond to Heidegger's criticism of (Cartesian) subjectivism. The idea of engagement is meant to imply that consciousness structures the subject and object at the same time. Note the parallel between this idea and James's proto-Husserlian phenomenology. This continuity alone ought to be enough to obviate the cognitive-affective debate.
28. Solomon, "Emotions, Thoughts, and Feelings," 87.
29. Ibid.
30. Alison Jagger, "Love and Knowledge: Emotion in Feminist Epistemology," in *Gender, Body, Knowledge*, ed. Alison Jagger and Susan R. Bordo (New Jersey: Rutgers University Press, 1989); Nussbaum, *Upheavals of Thought*, 26; Prinz, *Gut Reactions*, 76; and LeDoux, *The Emotional Brain*, 19, 69.
31. We are no longer using the term "affect" to refer to physiological sensations.
32. We should start a discussion of Kant's theory of emotion with his theory of feeling. Many commentators start with his comments about "affects" and "passions," but they fail to notice that affects and passions are two different sorts of mental entities. Affects are feelings, and passions are desires. I discuss passions in chapter 6, and in chapter 4 I give more reason to think that emotions are only secondarily related to desires. For a good argument that we should uncouple our analysis of Kant's theory of emotion from his notion of inclination, see Ido Geiger, "Rational Feelings and Moral Agency," *Kantian Review*, 16 (2011): 283–308. In addition, when we understand the nature of Kantian (intellectual) feelings, namely, that they are always the feeling *of* some cognitive event, we can better understand their relationship to motivation—as Geiger describes the feeling of respect: "It follows the dictates of reason, but it cannot itself formulate rational directives" (291). After we realize that affects, in Kant's sense, are feelings, we can characterize feelings in many different ways in relationship to their varying cognitive causes: moral and immoral feelings, strong and weak feelings, rational and irrational feelings, and so on.
33. The "Introduction" to the "Doctrine of Right" is actually an "Introduction" to the entire *Metaphysics of Morals*.
34. Here I am not referring to Kant's theory of passion (*Leidenshaft*), which will be addressed in chapter 6. Maria Borges is right to hold that Kant's

theory of emotion, as it takes emotions to be both cognitive and physiological, contributes a considerable amount to contemporary debates, but her characterization of the nature of this conjunction is not entirely accurate. Affects are feelings, for Kant, and feelings are *caused by* cognitive events; they are not themselves cognitive. While it is possible to put Kant's notion of passion "under the label of emotions," as Borges seems to, Kant distinguishes between passions and feelings (153). I argue in chapter 6 that since passions for Kant are mostly vices, the important question is how *not* to let affects turn in to passions. See Maria Borges, "What Can Kant Teach Us About Emotions?" *Journal of Philosophy*, 101 (3) (March 2004): 140–158.

35. Kant provides an important insight on the difference between his and Descartes's approach to the study of human nature. Descartes, on one hand, is largely concerned with a physiological, neurological study of humans; Kant, on the other hand, considers humans from the point of view of their free, mental lives: "[Descartes's] physiological knowledge of man investigates what nature makes of him: pragmatic [knowledge investigates] what man makes of himself" (A 119). For more on this distinction, see Robert Louden, "Anthropology from a Kantian Point of View: Toward a Cosmopolitan Conception of Human Nature," in *Rethinking Kant*, ed. Pablo Muchnik (Cambridge, UK: Cambridge Scholars, 2009).
36. Edmund T. Rolls, *Emotion Explained* (New York: Oxford University Press, 2007), 21. I find this textbook incredibly strange, even more so because it does not even attempt to argue that one can "explain" emotion without referring to any emotion, but instead referring to affective states that even the author does not call "emotions." It is one thing to attempt to replace ordinary language with more precise theoretical definitions or make the case that people are often confused in their experience; it is quite another thing to simply swap one concept for another, under the guise of rejecting "folk" theories of emotion. (See also LeDoux, *The Emotional Brain*, 16.) Rolls makes a distinction between emotions, which are initiated by stimuli in the external environment, and affective states, which are caused by a change in the "internal milieu," where hunger is an example of the latter and sadness is an example of the former. Furthermore, he believes that this difference is not sufficiently alienating, making it reasonable to study one and draw conclusions about the other. Clearly, Rolls's strategy for explaining emotions "scientifically" is to cut out their cognitive content in order to explain them biologically. This begs the question for the justification of an affective approach.
37. Jenefer Robinson, "Startle," *Journal of Philosophy*, 92 (1995): 53–57.
38. Ibid., 53: Robinson singles out Taylor's and Greenspan's work.
39. Ibid., 62.
40. Paul Ekman, "Expression and the Nature of Emotion," in *Approaches to Emotion*, ed. Klaus R. Sherer and Paul Ekman (Hillsdale, NJ: Lawrence Erlbaum, 1984), 329.

41. My point is that this argument about the way we should define emotion, if it boils down to privileging some instances of emotion and excluding others, does not constitute an interesting debate. The first and simplest step we can take in overcoming the cognitive-affective debate involves refusing to squabble over which emotions should or should not count as an "emotion." There can be no rational debate about how to cut the psychological cake. The only type of reason that can be offered in this debate comes from observations about the way we use the word "emotion," but contradictory reasons about the way that we *should* use the word "emotion" count too. It is common to point out that the current use of the word "emotion" is a relatively late linguistic development and that the term "passion," with its connection to passivity, is historically more prevalent. "Emotion," on the contrary, is formed from an active verb. These etymological musings are ultimately inconclusive: perhaps the recent linguistic development represents progress in the latent theory of emotion. See James Averill, "Emotion and Anxiety: Sociocultural, Biological, and Psychological Determinants," and Amélie Rorty, "Explaining Emotions," in *Explaining Emotions*, ed. Amélie Rorty (Oakland: University of California Press), 1980.
42. LeDoux, *The Emotional Brain*, 148.
43. The leap away from something one mistakenly took to be a spider might still work to confirm the affective theorist's notion of a necessary connection between perception and reaction, if the response is truly *universal.* Nevertheless, when the object of stimulation is more complex than a black dot, the ability to call the mental event a "perception" becomes strained.
44. Just as Hume uses skepticism to advance empiricism, James can sometimes appear to be a biological reductionist, but this polemic is in the service of advancing a psychological monism and moral naturalism. Calling himself a radical empiricist, James is suspicious of those thought processes that take themselves to be "pure." Instead, the most important intellectual truths are intuitive, and a product of our natural, psychological engagement with the world. It is not hard to see that James's pragmatism represents a dissatisfaction with the primacy given to reason in the history of philosophy, and so one defense proffered for the "lower" faculties comes in the form of a Romantic inversion of value. James seems to want to defend emotion from reason, and his strategy for doing this is to assert that emotions are more rational than reasons. This move is certainly not novel, and it is enjoying popularity currently. In fact, if we phrase the debate in these terms, it seems that alliances are redrawn and many more cognitive approaches, like Nussbaum's, end up agreeing with James and Prinz. Prinz's assertion that emotions are embodied must be understood as the idea that moral theory must be embodied, or that the body must play a foundational role in moral theory, as we can see with his renewed moral sense theory; Jesse Prinz, *The Emotional Construction of Morals* (New York: Oxford University Press, 2011).

45. Here we see an example of the supposed value of "living in the moment." Goodman, Russell, "William James", *The Stanford Encyclopedia of Philosophy* (Winter 2013 Edition), Edward N. Zalta (ed.), http://plato.stanford.edu/archives/win2013/entries/james/.
46. Hookway, Christopher, "Pragmatism", *The Stanford Encyclopedia of Philosophy* (Winter 2013 Edition), Edward N. Zalta (ed.), http://plato.stanford.edu/archives/win2013/entries/pragmatism/.
47. Rock believes that James is open to the criticism that he fails to make a distinction between perception and recognition, but this fact merely illuminates James's theory of perception. Irwin Rock, "A Look Back at William James's Theory of Perception," in *Reflections on the Principles of Psychology: William James After a Century*, ed. Michael G. Johnson and Tracy B Henley (Hillsdale, NJ: Lawrence Erlbaum Associates, 1990).
48. W. V. O. Quine, "Two Dogmas of Empiricism," *Philosophical Review* (60) (1951): 20–43; reprinted in his *From a Logical Point of View* (Cambridge, MA: Harvard University Press, 1953).
49. Lazarus, *Emotion and Adaptation.*
50. Prinz, *Gut Reactions*, 198.
51. Prinz flatly denies this, but I do not see what grounds he has for doing so. According to Prinz, an emotion is a perception of a core relational theme (like loss), and not individual objects or occasions (such as a dead body); still, are not core relational themes conceptual? Prinz, *Gut Reactions*, 50.
52. Prinz, "Embodied Emotions," in *What Is an Emotion?* ed. Robert Solomon and Cheshire Calhoun (New York: Oxford University Press, 1984), 57.
53. The burden of proof is on the evolutionary psychologist to find the neurological bases that would prove that certain emotional responses are purely physical.
54. Solomon, "Emotions, Thoughts, and Feelings," 79. This misunderstanding about the debate might be the reason that cognitive theories appear irksome to affective theorists. To the affect theorist, the cognitive theorist is obstinately asserting that people are somehow consciously and antecedently aware of the judgments that contribute to an emotion. They are right to object that such could not allow for the quickness of emotions, nor does it ring true from experience. Thus, affective theorists often caricature the cognitive approach as arguing that all emotions are like false, affected emotion. (See Goleman and the discussion in chapter 8.) In reality, it is not at all an important feature of a cognitive theory that the subject be consciously aware of the beliefs and judgments that are related to an emotion. What is important to the cognitive theorist is that the emotion be cognitively analyzable. In this way, emotions can be shown to be related to beliefs and judgments after the fact, and their genesis can be constructed into an intelligible psychic narrative, even if their causes remain unconscious psychic mechanisms.
55. In an attempt to bring together cognitive and affective approaches, Singer and Schachter have developed a "two components" theory, arguing that

emotions involve both physiological responses and their cognitive evaluations. S. Schachter and J. Singer, "Cognitive, Social, and Physiological Determinants of Emotional State," *Psychological Review,* 69 (1962): 379–399. As we have seen, neither side denies that emotions involve both cognitive and affective aspects. Nevertheless, the disagreement is likely about the relative importance of each aspect and the connotations of each.

56. Prinz, *Gut Reactions*, 236.
57. "[A]n incentive can determine the will to an action only insofar as the individual has *incorporated* it into his maxim" (R 6:24, emphasis mine); this is referred to as Kant's "Incorporation Thesis." See Henry Allison, *Kant's Theory of Freedom* (New York: Cambridge University Press, 1990).
58. Whether or not our minds are radically free and could possibly direct the processes of our brains without themselves being determined by the processes of our brains is a murky philosophical question, about which, to quote Socrates on his deathbed, I am in danger of not having a philosophical attitude. For a fuller discussion of brain processing, see "Conclusion."
59. A more thorough discussion of the difference between psychological and biological endeavors can be found in the "Conclusion" of this book.
60. Prinz, *Gut Reactions*, 180. For Prinz's account of mood, see pp. 187–188. Evidence that his rejection of cognitive theories of emotion stems from ignoring attitudinal emotions comes in the following: "[A]ttitudinal emotions contain both embodied appraisals...and representations of objects or states of affairs. These two components are bound together in the mind. The actual nature of the binding has not been adequately investigated" (181).
61. Nussbaum, *Upheavals of Thought*, 73.
62. My point here is not unlike Peter Goldie's insistence that we must consider the narrative structure of an emotion. See Peter Goldie, *The Emotions: A Philosophical Exploration* (Oxford: Oxford University Press, 2002).
63. On p. 264 of the *Anthropolgy*, Kant writes: "If only he has understanding and great power of imagination, an imposter who is himself unmoved can often stir others more by an affected (simulated) emotional agitation than by the real one. In the presence of his beloved, the man who is seriously in love is embarrassed, awkward, and not very attractive. But a man who merely pretends to be in love and has talent can play *his role* so naturally that he succeeds completely in trapping the poor girl he dupes, just because his heart is unbiased and his head is clear, and he is therefore in full possession of the free use of his skill and power to imitate very naturally the appearance of a lover."
64. How does this humor relate to emotion? Is it an example of forgiveness? Compassion for myself?

65. The goal of therapeutic self-understanding necessarily sees emotions as fluid and indistinct since it is familiar with the relationship between emotions and the ways that one emotion can turn into or reveal itself to contain another. Neu, a representative of a more cognitive approach, argues that the emotions do not qualify as natural kinds. He argues that emotions are determined by thoughts and so are too numerous to classify in only the most general groupings. See Jerome Neu, *A Tear Is an Intellectual Thing* (New York: Oxford University Press, 2002). Rorty concurs and adds that emotions cannot be "sharply distinguished from moods, motives, attitudes, [and] character traits." See Amélie Rorty, "Introduction," in *Explaining Emotion* (Oakland: University of California Press, 1980). Again, in this difference between the two approaches, we do not see a disagreement about the facts about emotion—Frijda's "laws" of emotion are not wrong, they are simply very general and open to the criticism of tautology—but a difference in focus and goal. See Nico Frijda, *The Laws of Emotion* (London: Lawrence Erlbaum, 2007); M. D. Lewis, "Self Organizing Cognitive Appraisals," *Cognition and Emotion*, 10, 1996: 1–26; and "Bridging Emotion Theory and Neurobiology through Dynamic System Modeling," *Behavioral and Brain Science*, 28, 2005: 105–131.
66. See Antonio Damasio, *Looking for Spinoza: Joy, Sorrow, and the Feeling Brain* (New York: Mariner Books, 2003).
67. See Martin Heidegger, *Being and Time*, trans. John Macquarrie and Edward Robinson (New York: Harper Perennial, 1962), 172.
68. The first two sentences of Book One of his *Treatise of Human Nature* runs thus: "All the perceptions of the human mind resolve themselves into two distinct kinds: impressions and ideas. The difference betwixt these consists in the degrees of force and liveliness with which they strike upon the mind." See David Hume, *Treatise of Human Nature* (Oxford: Oxford Clarendon Press, 1888).
69. Paul Ekman, "Basic Emotions," in *Handbook of Cognition and Emotion*, ed. T. Dalgleish and M. Power (Sussex, UK: John Wiley & Sons, 1999).
70. Paul Ekman, "a Methodological Discussion of Nonverbal Behavior," *Journal of Psychology*, 43 (1957): 141–149.
71. A possible objection might be that I am watering down an account and discussion of emotion. Perhaps. We should nevertheless remember that these bits of nonverbal communication can be important morally. Imagine, for example, a doctor who seeks informed consent but who ignores nonverbal communication, like the expression of misunderstanding or the body language associated with depression.
72. I am indebted to the work of Fanny Söderbäck for this reference.
73. Davidson's argument that global skepticism is impossible comes to mind.
74. Mary Carruthers, *The Craft of Thought: Meditation, Rhetoric, and the Making of Images, 400–1200* (New York: Cambridge University Press, 1998).

75. This characterization of consciousness is similar to Damasio's; Antonio Damasio, *The Feeling of What Happens: Body and Emotion in the Making of Consciousness* (New York: Mariner Books, 2000).
76. As we will see in chapters 4 and 6, Kant's comments about desire are related to, but not the same as, his theory of emotion. Kant takes desire, in a pejorative sense, to be related more to physical pleasure and pain than to intellectual pleasure and pain.
77. Prinz, *Gut Reactions*, 10.
78. Ibid., 4.
79. Cheshire Calhoun, "Cognitive Emotions?" in *What Is an Emotion?* ed. Robert Solomon and Cheshire Calhoun (New York: Oxford University Press, 1984), 328. Much of Calhoun's argument seems to be preceded by Amélie Rorty, "Explaining Emotions," *Journal of Philosophy*, 75 (3) (1978): 139–161.
80. Calhoun details the difference between what she calls "intellectual" or "evidential" beliefs and "experiential" beliefs to help explain emotion-belief conflicts. Experiential beliefs, which come from some kind of biased history, can intrude on one's intellectual beliefs like a kind of illusion. In this way, we can deny the intellectual validity of our emotions. Emotions, thereby, involve epistemic normativity: our emotions should match our intellectual beliefs. Calhoun concludes that emotions must be analyzed in terms of one's elaborate system of beliefs, which include "interpretive 'seeings as…' and their background cognitive sets" (342). She concludes that emotions are not beliefs, but interpretations, but this does not address her original criticism that emotions and beliefs, and now interpretations, are logically and ontologically distinct sets. Sherman argues that Aristotle's notion of *phantasia* offers this same insight about emotion. See Nancy Sherman, *Making a Necessity of Virtue: Aristotle and Kant on Virtue* (New York: Cambridge University Press, 1997), 61.

3 Emotions, Decision Making, and Morality: Evaluating Emotions

1. Patricia Churchland, *Braintrust: What Neuroscience Tells Us about Morality* (Princeton, NJ: Princeton University Press, 2011), 3.
2. Ibid., 24–25.
3. Ibid., 165. Note that here she uses the term "moral" in its normal sense of referring not just to caring behavior but to an analysis of that which is good.
4. Ibid., 13. She also highlights the role that mirror neurons play in unconscious mimicry in order to explain group conformity and empathy (145–162). This line of explanation further undermines the value of difference and dissent.
5. Joshua Greene, *Moral Tribes* (New York: Penguin Press, 2013), 25.

6. Nussbaum makes this point; see Martha Nussbaum, *Upheavals of Thought: The Intelligence of Emotions* (New York: Cambridge University Press, 2001), 8.
7. Iris Murdoch, *The Sovereignty of the Good* (New York: Routledge, 2001); Martha Nussbaum, *Upheavals of Thought: The Intelligence of Emotions* (New York: Cambridge University Press, 2001). Margaret Urban Walker, "Moral Understandings: Alternative Epistemology for a Feminist Ethics," *Hypatia*, 4 (2) (1989): 15–28. Lawrence Blum, *Friendship, Altruism, and Morality* (New York: Routledge & Kegan Paul, 1980).
8. Michael Stocker and Elizabeth Hegeman, *Valuing Emotions* (New York: Cambridge University Press, 1996), 82.
9. Ibid., 85.
10. Ibid., 137.
11. Ibid., 188–189.
12. Ibid., 202.
13. Solomon might be taken as one of the only remaining critics of emotion because he continues to highlight the importance of making conscious choices about emotions, to have a "willingness to become self-aware, to search out, and challenge the normative judgments embedded in every emotional response"; Robert Solomon, "Emotions and Choice," in *Explaining Emotions*, ed. Amélie Oksenberg Rorty (Oakland: University of California Press, 1980), 325. Following Sartre, he argues that, since "normative judgments can be changed through influence, argument, and evidence, and since I can go about on my own seeking influence, provoking argument and looking for evidence, I am as responsible for my emotions as I am for the judgments I make" (316). Solomon's thesis might be better expressed with the more general idea that we are responsible for ourselves. There seems to be no reason for him to focus on emotions since he does not have an example of someone who argues that we are totally passive with regard to the passions, but, rather, he is inspired by Sartre's general emphasis on personal responsibility. I agree with him in this regard. Nevertheless, the reactive battle cry that emotions are rational must not be premised on a deficient notion of reason. Solomon joins with the Romantic side, straw-manning Kant and the notion of reason, by arguing that the insight that emotions are values "wreaks havoc on several long cherished philosophical theses" like the idea that morality must be based on reason (313). We shall soon see that this is not true.
14. See Antonio Damasio, *Looking for Spinoza: Joy, Sorrow, and the Feeling Brain* (San Diego, CA: Harcourt Books, 2003); Antonio Damasio, *The Feeling of What Happens: The Body and Emotion in the Making of Consciousness* (Wilmington, DE: Mariner Books, 2000).
15. Damasio, *Looking for Spinoza*, 141.
16. Ronald de Sousa, "The Rationality of Emotions," in *Explaining Emotions*, ed. Amélie Rorty (Oakland: University of California Press, 1980); Ronald de Sousa, *The Rationality of Emotions* (Cambridge: MIT Press, 1990).

Perhaps the latter idea—that emotions break stalemates in rational thinking, like arbitrary whims—should count as a reason that emotions are *irrational.*

17. This theory is similar to de Sousa's "New Biological Hypothesis" of emotion, one and two; de Sousa, *The Rationality of Emotions*, 195–201.
18. Damasio, *The Feeling of What Happens*, 41.
19. Like other functionalists, Damasio ignores the fact that emotions do not always appear to be functional. In fact, there is reason to think that we more readily associate the emotions that do not appear functional with the term "emotion" than we do the mere triggering of an "emotional memory," as with remembering that there was a negative consequence associated with certain previous behaviors.
20. Bechara et al., "Deciding Advantageously Before Knowing the Advantageous Strategy," *Science*, 275 (1997): 1293–1294.
21. Damasio, *A Feeling of What Happens*, 163.
22. Ibid., 169.
23. Roxanne Khamsi, "Impaired Emotional Processing Affects Moral Judgments," *NewScientist.com*, March 2007.
24. Consider fictional accounts such as *The Clockwork Orange* or *The Eternal Sunshine of the Spotless Mind.* The real-world pursuit of treating neurobiological ailments, especially the post-traumatic stress disorder of war-traumatized soldiers, offers parallels.
25. Patricia S. Greenspan, *Emotions and Reasons: An Inquiry into Emotional Justification* (New York: Routledge, 1988).
26. Richard Lazarus, *Emotion and Adaptation* (New York: Oxford Press, 1991).
27. An affective theorist might suggest that some emotions, like the affection that yields a hug, are spontaneous and do not involve any such explicit thought. We should recognize instead that all chosen behaviors involve thought, and only a deficient understanding of cognition takes it to be slow and merely theoretical.
28. The argument that emotions are "modes of attention" that "track morally relevant news" strikes me as feeble (Sherman, *Making a Necessity of Virtue*, 39). It is comparable to responding to skepticism by praising the wonders of sight. First, it talks past the criticism of emotion. Second, it inadvertently reinforces the assumption that something like robot experience is possible. Third, it seems silly—like something out of Aristophanes's *The Clouds*—to praise something that is so much a part of normal experience.
29. De Sousa, "The Rationality of Emotions," 149.
30. Jesse Prinz, *Gut Reaction: A Perceptual Theory of Emotion* (New York: Oxford University Press, 2004), 9.
31. Sigmund Freud, "On Narcissism," *The Freud Reader*, ed. Peter Gay (New York: W. W. Norton, 1995).
32. Sigmund Freud, *Group Psychology and the Analysis of the Ego* (New York: W. W. Norton, 1989), 81.

33. Recently a student of mine resisted my moral way of reasoning by saying that "everything has an opportunity cost." By this he meant that moral actions always have some disadvantage of taking away from our lives in some way, like costing us time, money, or pleasure. To this Kant would say that valuing time, money, or pleasure (pragmatic concerns for our own happiness) above the demands of morality is the definition of evil. It reflects very poorly on our culture that this truth is so hard for us to see.
34. The reasons that we cannot seek to perfect others are a bit harder to understand. Denis canvasses Kant's arguments for this conclusion and finds them unsatisfactory, so long as we avoid being paternalistic; see Lara Denis, *Moral Self-Regard: Duties to Oneself and Others* (New York: Garland, 2001), 143–157.
35. A more serious discussion of moral commitment, as well as its relationship to true self-esteem, can be found in chapter 7.
36. One might object that it is sometimes necessary to give commands and that they, in any case, are a part of ordinary speech. A parent often gives commands to his child. Nevertheless, the moral ideal is respect. In most cases, asserting one's position of power over another person, even a child, is demeaning and morally problematic. In some cases, perhaps in cases of danger, it is not demeaning, but helpful. We can see here that nonverbal communication and understanding bears on moral evaluation.
37. Barbara Herman, *The Practice of Moral Judgment* (Cambridge, MA: Harvard University Press, 1996).
38. Damasio believes that affective reasoning is distinctly based on memory of reward and punishment, but his card game is entirely a task of achieving reward and avoiding punishment, so, in effect, the lack of a control-group disallows such a conclusion. There is, in fact, no evidence that the nondeclarative knowledge draws from a different kind of evidence, such as opinion, prejudice, or personal memory, than conceptual reasoning does. Furthermore, it is not clear why conceptual knowledge would be insensitive to considerations of reward and punishment and why a separate decision-making process is necessary to accommodate these considerations: those people who had the hunch did not stop behaving advantageously as soon as they understood the reason for it.
39. It is possible that the latter moral demand promotes the psychological health of the individual through cultural dictate, since there seems to be something inherently difficult in experiencing negative emotions as well as recognizing the negative emotions of others.
40. De Sousa, *The Rationality of Emotion*, 45.
41. Theodor Adorno, *Aesthetic Theory* (Minneapolis: University of Minnesota Press, 1998).
42. I do not agree with Freud that some emotions must be repressed. Perhaps I can be criticized as Pollyanna-ish, but clearly my theory of emotion holds that repression is, for the large part, necessarily bad.

43. See Freud, *Group Psychology and the Analysis of the Ego,* and Sigmund Freud, *Beyond the Pleasure Principle* (New York: W. W. Norton, 1990).
44. Greene, *Moral Tribes,* 39–43.
45. This is Rachels' argument about psychological hedonism. See James Rachels, *The Elements of Moral Philosophy,* 5th ed. (New York: McGraw-Hill, 2006).
46. Paul Rasmussen, *The Quest to Feel Good* (New York: Routledge Press, 2010).
47. Kant often makes a similar point about the difficulty of knowing what will make us happy; he uses this observation as evidence that pragmatic reasons are not a priori and hence cannot be morally binding.

4 MORAL FEELINGS

1. Scott Raab, "Louis C.K.: ESQ + A." *Esquire* (May 2011).
2. See the "Introduction" for a brief overview of terminology or chapter 6 for an explanation of Kant's theory of "affects" and "passions." Kant's use of the term "affect" is unlike that of affective theory, discussed in the previous chapters.
3. I am grateful to Samantha Matherne for her helpful commentary on this chapter.
4. Hursthouse expresses the common, prejudiced misunderstanding of Kant's theory of emotion, arguing that Kant does not believe that emotions can be part of our rational nature, nor can they be morally significant. See Rosalind Hursthouse, *On Virtue Ethics* (New York: Oxford University Press, 1999), 109. Geiger as well as others have shown that this criticism results from conflating the faculty of feeling with Kant's more derogative notions of desire and inclination. See Ido Geiger, "Rational Feeling and Moral Agency," *Kantian Review*, 16 (2011): 283–308.
5. My emphasis on Kant's theory of virtue will seem backwards to some, but I agree with Allen Wood's conclusion that the *Metaphysics of Morals* is the most complete statement of Kant's moral theory and that attention to it will alter conclusions drawn exclusively from Kant's earlier works. It is therefore unfortunate that many critics of Kant are not familiar with this later text. See Allen Wood, "The Final Form of Kant's Practical Philosophy," in *Kant's Metaphysics of Morals*, ed. Mark Timmons (New York: Oxford University Press, 2002). Similarly, most scholarly understanding of Kant's theory of emotion sees it as largely derogatory. On the contrary, I see the notion of virtue as foundational in Kant's ethics and, because of this, stress the importance Kant places on moral emotions.
6. For a good overview of Kant's theory of virtue see Lara Denis, "Kant's Conception of Virtue," in *The Cambridge Companion to Kant and Modern Philosophy*," ed. Paul Guyer (New York: Cambridge University Press, 2006); and Anne Margaret Baxley, *Kant's Theory of Virtue* (New York: Cambridge University Press, 2010). Most of their exegetical work is presupposed here.

7. I address those feelings and obsessions that Kant considers more morally problematic (affects and passions) in chapter 6. Additionally, I argue there, with Sorensen, that Kant's notion of affect is not uniformly critical. See Kelly Sorensen, "Kant's Taxonomy of Emotion," *Kantian Review*, 6 (2002).
8. See note in her translation of the *Anthropology*, p. 131. Mary Gregor, trans., *Kant: Anthropology from a Pragmatic Point of View* (Amsterdam: Nijoff, 1974).
9. Pointing to a passage from the third *Critique*, Sorensen shows that Kant takes feeling to be a constant accompanying effect of human consciousness: "But the strongest statement of the necessity of this sensible feeling is in the third *Critique*. Kant says that specific feelings of pleasure have become so mixed with our cognition that we have forgotten they exist; but the pleasure of, for example, 'being able to grasp nature and the unity in its division into genera and species' is what 'alone makes possible the empirical concepts by means of which we cognize nature in terms of its particular laws.' '*Even the commonest experience would be impossible without it* (that is, this pleasure)'" (CJ 187). Susceptibility to sensible pleasure and pain, then, is a condition at the deepest root of human experience." See Sorensen, "Kant's Taxonomy of Emotion," 113.
10. "The capacity (receptivity) for receiving representations through the mode in which we are affected by objects, is entitled *sensibility*" (CR A 19, B 33). All sensations of our bodies, like the feeling of pain or hunger, are of objects; we can see that they should be classified this way because of the fact that they are in space. Kant discusses the experience of taste and color, for example, as subjective feelings that result from objective sensations (CR A 28). Kant describes inner sense as being like a sixth sense in which we experience and put together all of our other sensations.
11. Although Kant is referring here merely to perceptual ideas, not real concepts, I am not advancing this account as accurate developmental psychology, merely as evidence that Kant takes feelings to follow from cognitive events.
12. See, for example, DR 212.
13. There he calls the possibility of pure practical reason acting on its own accord "the philosopher's stone": "The understanding, obviously, can judge, but to give to this judgment a compelling force, to make it an incentive that can move the will to perform the action—this is the philosopher's stone!" (L 45—page numbers refer to Louis Infield's translation). This philosophical difficulty stems from his distinction between pathological causes and intellectual causes. He holds that these two causes might nevertheless need to work in cooperation to bring about a moral action because the bare thought of morality might be insufficiently motivating: "There are actions for which moral motives are not sufficient to produce moral goodness and for which pragmatic, or even pathological *causae impulsivae* are wanted in addition; but when considering the goodness of an action we are not concerned with that which moves us to that goodness, but merely with what constitutes the goodness in and

of itself" (L 18). Here Kant seems to be preempting Schiller's later criticism, namely that Kant's moral theory requires that we extirpate natural moral feeling.

14. Guyer argues that Kant maintains this theory of pathological determination into his later works and that it *develops* into the conviction that moral feeling is necessary for morality. Paul Guyer, *Kant and the Experience of Freedom* (New York: Cambridge University Press, 1996), 30 and 337. Guyer believes that Kant maintained the requirement of subjective determination throughout his mature philosophy and that the feeling of respect takes over this role from other moral feelings.
15. Beck concludes that "all determination of the will proceeds (a) from the representation of the possible action (b) through the feeling of pleasure or pain (through taking an interest in the action or its effect) (c) to the act. The aesthetic condition, the feeling, is either pathological or moral: the former if the pleasure precedes the representation of the law, the latter if it follows it and is, as it were, pleasure *in* the law." Lewis White Beck, *A Commentary on Kant's Critique of Practical Reason* (Chicago: University of Chicago Press, 1963), 224. For my own part, it seems strange to identify pleasure and desire in this way, and it seems that an interpretation that does not merge pleasure and desire is preferable since they are two different faculties.
16. A fuller discussion of the content of the moral law will follow in the next chapter. The reader can also refer back to the "Introduction." A quick reminder: The moral law states that we must not make an unfair (selfish) exception for ourselves in our principles; instead, we must respect both ourselves and others. Grounded also in this primary respect for people are the moral duties to promote our own perfection and the happiness of others.
17. Sometimes Kant limits the use of the term "moral feeling" to refer to our satisfaction at having acted morally, but most other times it refers to our feeling of respect for the moral law (DV 399). In what follows, I am going to take our evaluations of projected or past actions as very near the idea of conscience and discuss conscience as a type of moral feeling.
18. It is not uncommon for people to react with discomfort and aggression or other odd emotions, which are often strangely strong—strong enough to overpower moral reflection—when confronted with moral obligations. I am reminded of a recent incident made popular by internet condemnation where beach goers recorded themselves mocking a sick and stranded dolphin, appearing to have a great fun, instead of helping it.
19. The feeling of the sublime is similarly morally instructive because it reinforces the worth of human dignity. Unlike with beauty, sublimity can be connected to moral interest. Objects arouse the feeling of sublimity when they make us aware of our own worth and rational vocation. Guyer argues that the feeling of sublimity is very much like the feeling of respect, but the feeling of the sublime involves a subreption, so that

we project sublimity in the object instead of, I suppose, keeping it for ourselves. Guyer, *Kant and the Experience of Freedom*, 221.

20. Page numbers of the *Observations of the Feeling of the Beautiful and Sublime* refer to John Goldwaith's translation. In the second *Critique*, Kant warns against merely imitating noble and sublime actions because then the performance of such acts would not be based on principle, but here we see that the feeling of the sublime is the recognition of virtue based on principle (CprR 84–85).
21. Roberto Benigni's film *Life Is Beautiful* comes to mind.
22. We might say that this feeling of freedom is akin to feeling that one has superpowers, as one might need in, for example, jumping off a bridge to save people trapped in a sinking car.
23. For a cataloging of the similarities and differences between the feeling of respect and the feeling of sublimity see R. R. Clewis, *The Kantian Sublime and the Revelation of Freedom* (New York: Cambridge University Press, 2009). Clewis reminds us that enthusiasm can be sublime; it is "the idea of the good with affect" (CJ 272). As we will see in chapter 6, affects can be irrational or they can arise from reason but act irrationally. Enthusiasm is then a kind of moral fervor that may or may not be misled but perhaps does not proceed with as much strategic calculation as is required. Nevertheless, the fact that Kant calls it sublime is a testament to the importance of moral feeling.
24. See Sorensen, "Kant's Taxonomy of the Emotions," 119–120. I am also indebted to Sorensen for reference to the following passage.
25. Ibid., 122.
26. MacBeath argues that the feeling of respect is felt as an imperative and not just an inclination because it presupposes reason. He also calls Kant's theory of moral feeling "breathtakingly absurd" because he believes that it is a "fiction conjured up out of a defective view of rational action." Kant's view is supposedly defective because, according to MacBeath, reasons do not need feelings in order to be effective. A. Murray MacBeath, "Kant on Moral Feeling," *Kant-Studien* 64 (1973): 289 and 313–314. MacBeath risks reasserting a dichotomy between reason and emotion in advancing this conclusion. Kant does give the feeling of respect as part of the answer to the question, "how is pure reason practical?" Nevertheless, this question was poised for an answer from moral psychology. The question should not be understood as pointing to the ineffectual nature of reason, but only to its weakness relative to pragmatic incentives. The question asks: how is it that people would ever choose to follow the moral law when there is always a strong inclination pulling us in the direction of selfish benefit? The answer is: because we have moral feeling. We are psychologically convinced of the necessity of the moral imperative. Kant does not hold that all reasons need feelings in order to be effective; nevertheless, many more reasons might have feelings attached to them than we previously realized.

27. Note that here respect is not described as itself morally motivating, as it might be construed, but rather the accompaniment to moral motivation.
28. Guyer gives the compelling argument that Kant's moral theory relies on and is open to psychological insight. For example, the fact that the moral law inspires respect is a psychological fact, not a metaphysical fact. Guyer, *Kant and the Experience of Freedom*, 366.
29. William Sokoloff, "Kant and the Paradox of Respect," *American Journal of Political Science*, 45 (4) (2001): 769. See Mark Packer, "Kant on Desire and Moral Pleasure," *Journal of the History of Ideas*, 50 (3) (1989): 429–442, for the contrary argument that Kant's emphasis on respect shows that emotions must play a role in his theory of autonomy.
30. Sokoloff, "Kant and the Paradox of Respect," 777.
31. See note 7 of the Introduction.
32. Sokoloff defends the uniqueness of respect.
33. See Packer, "Kant on Desire and Moral Pleasure" for the argument that Kant's emphasis on respect shows that emotions must play a role in his theory of autonomy.
34. See chapter 6 for a discussion of self-esteem and other feelings associated with the pursuit of self-perfection.
35. We are not here referring to sexual desire; the love of parents for their children is a better example.
36. Kant is not referring to literal distance with this metaphor, as Baron assumes, but this reading is definitely the most natural. See Marcia Baron, "Love and Respect in the Doctrine of Virtue," in *Kant's Metaphysics of Morals*, ed. Mark Timmons (New York: Oxford University Press, 2002).
37. Support for the legitimacy of this extrapolation comes in a following passage when Kant likens respect to the negative duty to "not exalt oneself above others" (DV 449). Similarly, he writes that the duty of gratitude is the duty of respect for the benefactor and that a failure of gratitude is caused by resenting someone being "above oneself" (DV 454 and 458; see also 459).
38. Of course, this suggestion comes up in the discussion of charity and not in the discussion on private property because Kant would by no means justify theft.
39. Geiger writes that respect for the moral law, respect for humanity, love, and sympathy (as well as conscience and self-esteem) are all at base united because they all follow from proper comprehension of the moral law. See Geiger, "Rational Feelings and Moral Agency," 292.
40. This is the first sense of the term "emotional universalism": that we must recognize the universality and ubiquity of emotion and accept it. Indeed, it is repression and avoidance (of the "sickrooms and poorhouses") of these feelings that constitute one possible first step toward evil.
41. One might object that exposing oneself too often to people in need might also cause one to lose touch with our moral duties. This phenomenon is referred to as "compassion fatigue"; the idea is that one can become exhausted from helping, or even wanting to help, too much. I think that

this is an imprecise description of an actual experience. Instead, it is likely the case that the people who work in the helping professions, like police officers or counselors, become frustrated because they *cannot* help as much as they would like. Instead of suffering from wanting to help, the suffering truly comes from not being able to help. Once we set our minds to trying to help others, we soon realize that it is amazingly difficult to find something to do that will actually help. As Kant acknowledges, much of the human suffering in the world is the result of unjust political institutions, which are difficult to change. In the last chapter we will discuss the question of how much is enough, as well as self-esteem and self-satisfaction in the context of never-ending duties, but it is perhaps sufficient to say here that helping others and, most importantly, remaining open to helping others is not easy.

42. Beck discusses the fact that choosing duty over inclination is an illustration of morality, not morality itself. Lewis White Beck, *A Commentary on Kant's Critique of Practical Reason* (Carbondale: University of Chicago Press, 1960), 228.
43. Barbara Herman, *The Practice of Moral Judgment* (Cambridge, MA: Harvard University Press, 1996).
44. Although charting the changes in Kant's conception of moral motivation is beyond the scope of this project, let me be clear about my take on the topic of "over-determination." Henson argues that Kant should have accepted that both moral feelings and rational motivation together should be seen as morally acceptable motivations—he calls this "over-determination" because the action springs from (is determined by) more than one source. Herman argues that when moral feelings are present purely rational motivation must itself be sufficient in order for the action to count as virtuous. Guyer believes that both Henson and Herman are wrong because he thinks that Kant holds that all inclinations, emotions included, are subjected to a rational censor, and hence chosen. Therefore rational determination is never absent or insufficient. My focus here is on the importance of moral feeling. A proper understanding of what feeling is, namely the physical effect of a thought or experience, helps us to see both that actions are always overdetermined, whether by respect for the moral law or some other feeling or affect, and that feelings, like thoughts, must be evaluated as to their source and selfishness.
45. Guyer examines the *Metaphysics of Morals* for evidence that Kant believes that we have a duty to cultivate moral feeling. He argues that, first, the duty to outwardly conform to duty may require feelings to perform. Second, we have a duty to know ourselves and know whether or not we are motivated by duty. This requires psychological knowledge. Third, duty for duty's sake should usher in moral feelings. Fourth, duties of respect require we refrain from emotionally injuring others. Guyer, *Kant and the Experience of Freedom*, 382–384. Similarly, in the *Lectures on Ethics*, Kant argues that active obligations (*obligatione activa*) include the obligation to have a certain disposition, or character (*Gesinnung*).

46. Here we can see that he understands virtue to be something over which we have control, hence something regarding which we have a duty. Virtue ethicists would differ on this definition.
47. Geiger writes: "If we lacked [the feeling of respect], we would be incapable of moral action" (Geiger, "Rational Feelings and Moral Agency," 290). As odd as this seems, it is true.
48. In Baron's *Kantian Ethics Almost without Apology*, the lone apology is offered for Kant's deficient treatment of moral affect. Even though Baron, to my mind successfully, defends Kant from the charge of moral coldness, as I have also done here, she argues that Kant still sides with the Stoic against compassion, since "there cannot possibly be a duty to increase the ills in the world" as the sharing of the feeling of another's suffering would suggest (DV 457). To continue to fault Kant on this score, after everything we have seen, is, I think, a failure of philosophical flexibility, especially since, directly after this remark, Kant writes: "But while it is not in itself a duty to share the sufferings (as well the joys of others), it is duty to sympathize actively in their fate; and to this end it is therefore an indirect duty to cultivate the compassionate natural (aesthetic) feelings in us, and to make use of them as so many means to sympathy based on moral principles and the feeling appropriate to them" (DV 457). Marcia Baron, *Kantian Ethics Almost without Apology* (Ithaca, NY: Cornell University Press, 1999)
49. For a fuller account of beauty as a symbol of the good, see Henry Allison, *Kant's Theory of Taste, A Reading of the Critique of Aesthetic Judgment* (New York: Cambridge University Press, 2001).
50. Iris Murdoch, "The Sovereignty of the Good over Other Concepts," in *The Sovereignty of the Good* (New York: Routledge, 1971), 82.
51. Ibid., 85.
52. In his guidebook to Kant's philosophy, Guyer enumerates six connections between aesthetics and ethics, some of which overlap with our discussion here: (1) "aesthetic experience can present morally significant ideas in an imaginative and pleasing way"; (2) "the experience of the dynamical sublime so centrally involves the intimation of our capacity to be moral"; (3) "there are significant parallels between our experience of beauty and the structure of morality"; (4) "in the experience of beauty we can actually feel that the world is consistent with our aims, including our ultimate moral aim"; (5) quoting Kant, "the beautiful prepares us to love something...without interest" (CJ, general remark following §29); and (6) "the cultivation and realization of common standards of taste in a society can be conducive to the realization of...'lawful sociability.'" Paul Guyer, *Kant* (New York: Routledge, 2006), 324–328.
53. Geiger, "Rational Feelings and Moral Agency," 283.
54. This does seem the normal age at which children begin to experience "happy tears." I cannot avoid speculating about the cause of this development: Perhaps at this age we are also able to better grasp disinterested concepts, as well as a deeper sense of human mortality.

55. Even if my care about the mouse's reunion with his family was based in my own desire to then experience sympathetic happiness, although it was not, this feeling would still count as a moral feeling because the fact that I experience sympathetic happiness is a good thing. In general, hedonism falls apart as a theory when it runs up against natural moral feelings.
56. Let us tarry a moment to meditate on crying. Why do we cry when we are happy? I am not sure what kind of answer I could possibly be looking for with such a question. The question "Why do we cry when we're sad" has, perhaps, a different sort of answer—perhaps the sort that Darwin gives about functionality. In *The Expression of the Emotions in Man and Animals*, Darwin guesses that emotional crying is related to the normal function of tears to moisten the eye. Nevertheless, we should not expect too much from evolutionary functionalism's brand of just-so stories, and the type of cause we are looking for need not be a final cause, that is, the purpose of crying. (See Charles Darwin, *The Expression of the Emotions in Man and Animals* [New York: D. Appleton, 1873], 163.) Such a hypothesis is nonfalsifiable. Instead, we can wonder if there is any similarity between the tears of happiness and the tears of sadness. It seems, on the face of it, that there should be. Although it is often assumed that tears are cathartic, that is, they facilitate a clearing of the mind of getting done with a particular thought, they can also have the opposite effect and accompany increased brooding (see Benedict Carey, "The Muddled Tracks of All Those Tears," in *The New York Times*, February 2, 2009). We must also remember that crying can have the effect of eliciting sympathy. We might sometimes need to be suspicious of ourselves when we cry. Neither of my examples elicits such a suspicion. Particular thoughts or perceptions trigger tears, and just as crying is sometimes a sort of "breaking down" whereby we cannot accomplish other tasks, the thoughts that trigger crying are often of human mortality and frailty. Perhaps even the successes of life remind us how simple, hence mortal, we as humans are. It takes a certain amount of courage to be emotional and to cry with happiness (and perhaps even to truly be happy) because it is an acknowledgment of our dependency or finitude. As we saw previously, the feeling of sublimity that accompanies the experience of morality is one wherein the self is both strengthened and overcome. Perhaps we cry because we feel overpowered by the truth—the truth of our own mortality—and find strength and happiness in the face of it.
57. Perhaps I sympathize with all living things or perhaps I only sympathize with cute baby animals... or perhaps I have been trained to sympathize with all animals because of the number of animated animal movies I saw as a child. We will leave the authenticity of *my* feeling out of the example.
58. This fact gives us insight into Kant's remark that we do not have a duty to, by feeling more and more sympathy, increase the suffering in the world (DV 457). Nevertheless, not feeling sympathy would most likely be worse.

59. Most people are unaware—perhaps intentionally so—that American animal agriculture involves heinous instances of unnecessary suffering; they picture instead the traditional image, with which we have all been brought up, of a farmer raising animals in a pasture.
60. For a similar account of psychologically contextualized moral reasoning, see Herman, *The Practice of Moral Judgment.*
61. For a critical account of the duties to avoid sexual self-degradation, see Lara Denis, *Moral Self-Regard: Duties to Oneself in Kant's Moral Theory* (New York: Garland, 2001), 102–123.
62. Kelly Sorensen raised the questions of whether "emotion can correct and identify bad reasoning. [and] How much can emotion push reason?" This question is related to whether or not we should truly regard feelings as "blind" or whether they might sometimes lead the way to help us discover true moral principles or correct moral choices. I venture that many of us have had experiences of the latter. The best way that we can account for this experience is with a theory of unconscious thought—so that the feelings are caused by thoughts of which we are not totally aware; see my "Kant's Theory of the Unconscious" Available at: https://courses.cit.cornell.edu/ac385/UNYWEMPwebsite3/Diane%20Williamson,%20Kants%20Theory%20of%20the%20Unconscious.doc.
63. Kant holds that we should not promote the immoral ends of others (DV 388).
64. Kant remarks that we must, to have a virtuous character, find a way to make complying with positive duties as automatic and necessary as complying with negative duties (DV 390).
65. This explanation of feelings is helpful because, while it recognizes that feelings can be genuinely insightful, it does not paint this insight as some kind of magical voice entirely separated from our rational minds.
66. Kant writes that we can be wrong about whether or not something is our duty but we cannot be wrong about whether or not *we have judged* an act to be good or bad. To my mind, this says very little and does not truly imply that conscience is infallible (DV 401).
67. Our discussion of Kant's notion of virtue leads us to the question of his relationship to virtue ethics. I have commented on some similarities and differences between Kant and Aristotle throughout, as I will continue to, but, as I remarked earlier, I hold that the most commonplace understanding of Kantian moral theory, exactly because it largely ignores the importance of virtue and moral feeling, is largely mistaken. Betzler notes that virtue ethics focuses on character, the ideal of human flourishing, and lists of virtues as perfections of human natural capacities, and she distances Kant from virtue ethics proper because she holds that the popular view of him is that he remains faithful to "the priority of the right." Monika Betzler, "Kant's Ethics of Virtue: An Introduction," in *Kant's Ethics of Virtue*, ed. Monika Betzler (Boston, MA: Walter de Gruyter, 2008), 26. In truth, Kant has a better understanding of virtue than Betzler realizes. His notion of virtue is as a commitment to (rational)

moral principles. Being able to grasp the rational content of moral value and to evaluate things and prioritize goals thereby is the basis of morality. Simply not doing something wrong—the right in absence of the good—does not have any moral value in itself.

68. Julia Annas, *The Morality of Happiness* (New York: Oxford University Press, 1993), 53.
69. Henry Sidgwick, *The Methods of Ethics* (Indianapolis: Hackett, 1981); Annas, *The Morality of Happiness*, 56.
70. As Baron notes, in English, both the words "duty" and "obligation" carry a negative connotation. Baron speculates that for many Americans the term duty is associated with military duty. Baron, *Kantian Ethics Almost without Apology*, 16. It is hard to imagine a use of the word "duty" that is associated with something we actually *want* to do. Kant is well known for contrasting duty and inclination, and it is this contrast that calls into suspicion his insistence that the notion of duty must be at the heart of moral theory. As Paton points out, "[I]n the very idea of duty there is the thought of desires and inclinations to be overcome." H. J Paton, *The Categorical Imperative: A Study in Kant's Moral Philosophy* (Philadelphia: University of Pennsylvania Press), 46. Kant's definition of virtue similarly implies this kind of internal conflict. Yet, Auxter points out that the German *Verbindlichkeit* (obligation) carries a more positive sense of boundedness, as "moral...activity is the basis for the tie we feel with others." Thomas Auxter, *Kant's Moral Teleology* (Macon, GA: Mercer University Press, 1982), 163–164.
71. The idea that we should mimic the virtuous person threatens to make virtue irrational. For an example see Keiran Setiya, *Reasons without Rationalism* (Princeton, NJ: Princeton University Press, 2010).
72. Lest we find the constant threat of backsliding into mere self-control or the assumption that emotions have a miraculous parallel with virtue; Annas, *The Morality of Happiness*, 67.
73. Annas, *The Morality of Happiness*, 53. Confusing inclination and emotion leads to a defense that highlights the role of impurity, as opposed to the holy will, in Kant's ethics. See Robert B. Louden, *Kant's Impure Ethics: From Rational Beings to Human Beings* (Oxford, New York: Oxford University Press, 2002); and Nancy Sherman, *Making a Necessity of Virtue* (New York: Cambridge University Press, 1997). Gregor similarly refers to man as a "moral being with an animal nature"; Mary Gregor, *Laws of Freedom* (Oxford: Blackwell, 1963), 128. In this way, one might argue that Kant championed pure reason, but that he made concessions for the human case because we do not have divine wills and are thus a mixture of pure reason and inclination/emotion/feeling. This defense paints an overly austere picture of "pure reason," which need not and does not exclude feeling, as we have seen. (It is this mistake, confusing feeling with inclination, that leads Gregor to think that Kant would prefer holiness, which she describes as the lack of feeling, over virtue. There is no reason to think that holiness equates with the lack of feeling instead

of with the lack of temptation. In other words, there is no reason to think that Kant mistook angels for robots.) Gregor, *Laws of Freedom*, 175.

74. Furthermore, we must appreciate the context of Kant's characterization of virtue: he is explaining the way that the "Doctrine of Virtue" similarly involves constraint, as does the "Doctrine of Right" since "all duties involve a concept of constraint through a law" (DV 394). So, "what essentially distinguishes a duty of virtue from a duty of right is that external constraint to the latter kind of duty is morally possible, whereas the former is based only on free self-constraint" (DV 387). If it is in fact the case that Kant believes that "rational *natural*" beings are always tempted by pleasure, and that the moral law is not simply technically a constraint but also always *felt* as a constraint (because he draws attention to constraint in this context), he nevertheless emphasizes that it is a *self*-constraint, and therefore simultaneously voluntary and affirmed (DV 379). The relationship between duty and constraint is a topic that requires further inquiry. Kant writes that we do not have the duty to promote our own happiness only because we do so naturally and not "reluctantly" and so it is not a constraint (DV 386). It remains to be seen whether or not he means anything more than the idea that the notion of "duty" or "command" seems to require as content a behavior that we were not, necessarily, going to do already, just as we can only have as a duty something that is possible. If it is not merely a semantic point, it seems to have strange and unsettling consequences, such as something no longer being a *duty* when we *want* to do it. If it is merely a semantic point, the degree of argumentative weight it carries is unjustified. For a more complete discussion of Kant's notion of duty, see Baron, *Kantian Ethics Almost without Apology*.
75. Horn argues that, for Kant, the person who is perfectly good would have achieved a constant feeling of love—although that would still not be the basis of her actions. See Christoph Horn, "The Concept of Love in Kant's Virtue Ethics," in *Kant's Ethics of Virtue*, ed. Monika Betzler (Boston, MA: Walter De Gruyter, 2008), 155.
76. See Gregor, *Laws of Freedom*, 171.
77. Kant suggests earlier that "the Graces" represent the proper number of guests at a dinner party.
78. Borges, ignoring this direct reference to the Stoics, as well as a reference to the necessity of moral feeling in an earlier passage she cites (DV 410), concludes that the extirpation of emotions is a Kantian ideal. A more thorough reading of these passages helps us to make sense of Kant's comments about apathy and *tranquilitas*. See Maria Borges, "Physiology and the Controlling of Affects in Kant's Philosophy," *Kantian Review*, 13 (2) (2008).
79. In the *Anthropology*, as with many of his more lighthearted discussions, Kant discusses some of the more prosaic aspects of social enjoyment. Kant characterizes the dinner party as the highest ethicophysical good: "The good living which still seems to harmonize best with virtue is a

good meal in good company (and if possible with alternating companions)" (A §88). Obviously, this idea is not as important as the Highest Good of the second *Critique* (happiness only in accord with virtue), but it is still surprising to see him give such high praise to dinner parties. He discusses the ways that dinner party conversation complements and promotes philosophical thought: "Eating alone (*solipsimus convictorii*) is unhealthy for a philosophizing man of learning, it does not restore his powers but exhausts him...it turns into exhausting work and not the refreshing play of thoughts" (A §88). Kant values dinner parties so highly that he engages in many Emily Post–type recommendations for their success, concerning, for example, the proper number of guests and rules for successful conversation (A §88). In the "Doctrine of Virtue" Kant also explains that enjoying social interaction itself constitutes a virtue. There he makes a connection between social intercourse and his moral notion of cosmopolitanism: "[W]hile making oneself a fixed center of one's principles, one ought to regard this circle drawn around one as also forming part of an all inclusive circle of those who, in their disposition, are citizens of the world" (DV 473). He explains that the social virtues lead indirectly to an ideal world, or, we might say, the Kingdom of Ends. It is a duty of virtue to cultivate "a disposition of reciprocity— agreeableness, tolerance, mutual love and respect (affability and propriety, *humanitas aesthetica et decorum*)" (DV 473).

80. We can see that "feeling" might not be the best translation in the passage above.

5 EMOTIONAL UNIVERSALISM AND EMOTIONAL EGALITARIANISM

1. According to Paton (looking only at the *Groundwork*) there are five formulae of the categorical imperative: I. Formula of Universal Law; Ia. Formula of the Law of Nature; II. Formula of the End it Itself; III. Formula of Autonomy; and IIIa. Formula of the Kingdom of Ends. See H. J. Paton, *The Categorical Imperative* (Chicago: University of Chicago Press, 1948). In this chapter I will focus on I and II; I will also discuss Ia and IIIa. As we shall see, there is also reason for thinking, when looking in Kant's other works, that there are more than five formulae.
2. Kant intends the different formulations of the categorical imperative to all express the same underlying idea. On the one hand, this cannot be the case because they generate fundamentally different types of duties (perfect and imperfect) that do not overlap. On the other hand, the fact that there are different types of duties does not entail that one formulation of the categorical imperative cannot prescribe both. Confusion creeps in when the first formulation of the categorical imperative is split into two: the contradiction in conception test and the contradiction in willing test. This splitting as well as its ability to delineate between perfect and

imperfect duties is not successful. Nevertheless, it does not follow that the different versions of the categorical imperative are not equivalent. For a fuller defense of the equivalence of the various formulations of the categorical imperative, see Onora O'Neill, *Constructions of Reason* (Cambridge, UK: Cambridge University Press, 1990). Furthermore, it should be noted that I do not here attempt to explain or defend Kant's deduction (hence justification) of the categorical imperative, although I am partial to the argument that it is a "fact of reason," which has been discussed elsewhere. For a discussion of its strongest means of deduction, see Mark Thomas Walker, *Kant, Schopenhaur, and Morality: Recovering the Categorical Imperative* (New York: Palgrave Macmillan, 2012).

3. O'Neill (Nell) outlines these different types of testing for a contradiction. See Onora Nell, *Acting on Principle* (New York: Columbia University Press, 1975). If it turns out that a firm break between these two tests cannot be made in this way, no harm is done.
4. This idea is what is behind Hegel's "empty formalism" objection to Kant's overreliance on the notion of noncontradiction (paragraph 185 of *The Philosophy of Right*). For a fuller discussion of this criticism, see Sally Sedgwick, "Hegel's Critique of the Subjective Idealism of Kant's Ethics," in *Journal of the History of Philosophy*, 29 (1) (January 1988).
5. Herman argues that the first formulation cannot alone rule out murder. If we assume some moral anthropological knowledge, that is, that the agent willing has a finite body, I think it can. If murder makes willing impossible then it's a contradiction (in willing) for me to will murder. See "Murder and Mayhem," in Barbara Herman, *The Practice of Moral Judgment* (Cambridge, MA: Harvard University Press, 1993).
6. For another argument for the importance of the FLN formulation see Henry Allison, *Kant's Groundwork for the Metaphysics of Morals: A Commentary* (New York: Oxford University Press, 2011). Although it will not equalize the value of humans and animals, environmentalists might also find this version of the CI more fruitful.
7. Although, it may be the case that the two types of testing procedure cannot in fact be separated in this way, especially if we are to accept Kant's argument that all of the formulations of the CI express the same idea at base.
8. Louden argues against O'Neill that since Kant values the autonomy of reason so highly he could not possibly have established a procedure that commands specific actions and hence robs us of our own rational prerogative (p. 114). It is true that the content of specific positive duties cannot be determined ahead of time, but it seems that the content of negative duties is more uniform; although, empirical moral judgment will be required still to recognize our duties in either case. See Robert Louden, *Morality and Moral Theory* (New York: Cambridge University Press, 1992). For a good characterization of the empirical nature of moral judgment, see Barbara Herman, *The Practice of Moral Judgment.* Additionally, we should not forget that moral dilemmas are possible,

which Kant makes clear in a number of places, most notably with his inclusion of "casuistical questions" in the "Doctrine of Virtue" that conspicuously leave out the answers. By making morality a function of reason, Kant ensures that moral reasoners will always feel the responsibility to think for themselves and will not accept the conclusions of others secondhand. The categorical imperative is sufficient as a principle for guiding reflection; nevertheless, guiding reflection should not be misunderstood as prescribing specific actions. See also John Rawls, *Lectures on the History of Moral Philosophy* (Cambridge, MA: Harvard University Press, 2000), 148–149.

9. "Every human being has a legitimate claim to respect from his fellow human beings and is *in turn* bound to respect every other" (DV 462). "I cannot deny all respect to even a vicious man as a human being" (DV 463).
10. If the murderer has taken complete leave of his senses, then he might not require such respect and would even himself wish to be deterred by any means necessary.
11. Note that a more informative discussion about moral decision-making for a specific situation required reference to a number of different emotions: anger, calming down, fear, courage, and so on.
12. Jürgen Habermas, *Moral Consciousness and Communicative Action* (Cambridge: MIT Press, 2001), 120.
13. Emotions can easily be acceptable to all rational beings when they are universal emotions; they are not likely to be acceptable when they are confused or selfish.
14. James Rachels, *The Elements of Moral Philosophy*, 4th ed. (New York: McGraw-Hill College, 2002).
15. Hannah Arendt, *Lectures on Kant's Political Philosophy* (Chicago: University of Chicago Press, 1989).
16. Also see Robert Kane, *Through the Moral Maze* (New York: Parragon House, 1998) for such an interpretation.
17. Nevertheless, the CI, while similar, is not the same as the Golden Rule because people might actually consent to debasing themselves but the CI forces a rational, respectful position. While we are morally required to put in the effort to actually have the communicative back and forth, we also cannot accept the consent of another when it threatens to debase him.
18. Is the "transcendental formula of public right" a formulation of the CI? We can preliminarily note that the mere fact that it is a political principle does not prevent it from also being a moral principle since true politics and morality cannot be in conflict for Kant: "for true politics must bend the knee before right" (PP 125). Kant calls the transcendental principle of publicness an ethical and juridical principle, and the categorical imperative, as we see from the *Metaphysics of Morals*, is also both an ethical and juridical principle. We have a moral duty to promote perpetual peace; yet this end is co-guaranteed by Nature or Providence. As with Kant's regulatory

notion of the afterlife offered in support of our desire to believe in the reality of the Highest Good, he seeks to show the ways that progress towards Perpetual Peace is naturally attained in order to demonstrate to his reader that it is not a hopeless goal. Furthermore, Kant's "deduction" of TPP is incredibly similar to his deduction of the first formulation of the CI: he abstracts from all the material aspects of "public right" and is "left with the formal attribute of publicness" (PP 125). Recall Section One of the *Groundwork*, in which Kant begins with the formal idea of lawfulness as such in order to derive the moral idea of lawfulness. There we have the mere idea of lawfulness and here we have the mere idea of publicness. A law is essentially public, just as morality is essentially the government of all people, or individuals as they are rational beings.

19. Ibid., 17. Arendt does not consider that we might hold ourselves to a higher, not a lower, moral standard. In theory there would be nothing wrong with this, but, as we shall see with our discussion of the equation between self-respect and other-respect, there is something psychologically dangerous about it. Holding ourselves to a higher standard may imply either that we think we are better than others or that we think we are worse than others and must pay a penance. It may also imply that we are afraid to engage in moral discourse, not wanting to share or attempt to discover the inner moral life; such is a fear of intimacy.
20. Peter Singer, *The Expanding Circle: Ethics and Sociology* (New York: Farrar, Straus, & Giroux, 1981).
21. See chapter 7 for a discussion of moral perfectionism and moral consistency. We live in a world in which many of us commit immoral actions every day. Our understanding of morality cannot begin with the assumption that we are already morally perfect; instead we must accept that we are not perfect and work to improve, even if such improvement is extremely difficult. We cannot let the difficulty of achieving them water down our ideals.
22. It would be a mistake to conclude that Kant believes that we should share all of our thoughts with others. In the last paragraph of the *Anthropology*, Kant considers a hypothetical species in which there was no difference between thought and language—a species of people who said everything they thought. He concludes that humans could not live in peace under such conditions, "hence, it is part of the original composition of a human creature, and it belongs to the concept of the species, to explore the thoughts of others, but to conceal one's own" (A 332). Kant suffers no shortage of tips for polite dinner party conversation that confirm this requirement that we limit the disclosure of thought. Nonetheless, in the *Anthropology*, Kant goes on to write that this natural tendency to conceal leads to lying and that we must, as rational beings, combat this consequence. Therefore, it seems that Kant leaves the distinction between polite concealment and lying up to the rational subject.
23. In taking up this Kantian idea, Habermas is often criticized for his naiveté, and I am opening myself up to this same set of criticisms. Of course,

the goal of any given speech act can be a number of different things, especially the achievement of power over the other person, situation, or vis-à-vis the institution. I sympathize with Habermas and his political/psychological idealism, and I am here speaking normatively about our psychological/epistemological need for recognition on which other types of communication depend.

24. To what extent is it morally permissible to privilege ourselves and our loved ones? To the extent that everyone should privilege themselves and their loved ones. Many actions are owed to ourselves (as a part of our natural self-perfection), children (as a part of the duties of parents), or friends (as a part of the duties of friends). None of these requirements strike us as unfair or selfish, but if we threaten harm to another to fulfill these duties, we must curb our self-oriented duties and use objective moral reasoning to adjudicate the conflict of interests. Peter Singer also addresses this question in his chapter "One Community," in *One World* (New Haven, CT: Yale University Press, 2004).
25. See James Averill, "Emotion and Anxiety: Sociocultural, Biological, and Psychological Determinants," in *Explaining Emotion*, ed. Amélie Rorty (Oakland: University of California Press, 1980).
26. MM 212–213. We might agree that emotions do not tell us anything about objects, but does that mean that they cannot be universal, or that they do not tell us anything about the "our condition." It seems patently false that emotions and/or feelings do not tell us about ourselves and our nature. Therefore, we ought not interpret this clause in that way. Instead, it seems that by arguing that feelings cannot be connected to the "cognition of our condition" he means to contrast them to feelings that claim universality because they refer to the human condition, or cognition in general, such as aesthetic feeling. In the *Metaphysics of Morals* Kant also accepts that moral behavior yields moral feeling. Such a feeling should also claim universality, as the moral law is itself universal (see CJ 125).
27. Theory of Moral Sentiments: I.1.5 and II.1.5; Adam Smith, *The Theory of Moral Sentiments* (Salt Lake City: Gutenberg Press, 2011).
28. See Sigmund Freud, *The Psychopathology of Everyday Life*, trans. Anthea Bell (London: Penguin Books, 2002).
29. For such a criticism see Bernard Williams, *Moral Luck* (Cambridge, UK: Cambridge University Press, 1982).
30. Adorno and Horkheimer discuss this parallel in *Dialectic of Enlightenment* (New York: Continuum, 1976).
31. Humanity does not have inherent value because of its biological status. This confusion is perhaps made by opponents of abortion. If nonrational animals have rights, for Kant, it is only because of their relationship to rational beings. Children, for example, are the developmental precursors to rational beings, and so much be respected and reared as such. A Kantian environmental ethics is decidedly human-centered, but Kant nevertheless sees human moral value in the appreciation of nature.

32. It can also be misconstrued as deference because of one's station or achievements. While elders, as well as those who have achieved great things, deserve respect, this type of respect differs from respect for universal human dignity; we might call the former social respect.
33. Think about Kant's distinction when you hear politicians insist that the government should regulate individual's life choices before they can be allowed to be beneficiaries of government charity. Charity is only charitable if it is freely given and in accordance with the beneficiaries', not the benefactors', understanding of happiness.
34. See Murray Bowen, *Family Therapy in Clinical Practice* (New York: Jason Aronson, 1994).
35. See DV 469–470.
36. See Isaiah Berlin, "Two Concepts of Liberty," in *Liberty* (Oxford: Oxford University Press, 2002).
37. See Henry Allison, *Kant's Groundwork for the Metaphysics of Morals: A Commentary* (New York: Oxford University Press, 2011). This rights-based way of thinking about moral respect also has the consequence of making caring relationships (and the "private sphere") amoral.
38. Also see R 623 for further evidence that Kant believes that if everyone were virtuous the world would be a wonderful place.
39. See Habermas's argument for the priority of dialogical over monadological reason.
40. See also Denis on this topic; Lara Denis, *Moral Self-Regard: Duties to Oneself in Kant's Moral Theory* (New York: Routledge, 2001), 159 and 209.
41. For the complete argument that we have such a duty to ourselves, see Carol Hay, *Kantianism, Liberalism, and Feminism: Resisting Oppression* (New York: Palgrave MacMillan, 2013).
42. For an account of moral engagement as sensitivity and responsiveness, see Elise Springer, *Communicating Moral Concern: An Ethics of Critical Responsiveness* (Cambridge: MIT Press, 2013).

6 THE PATH OF VICE

1. I am grateful to Robert Gressis for his helpful comments on this chapter.
2. It is useful to realize that, by "anthropology," Kant means something much closer to what we mean by "psychology," that is, a study of human mental capacities and behaviors as they are common among humans. Kant defines psychology as the study of inner experiences as they fall under natural laws (A§7). Examples of such natural laws are the categories of the first *Critique*. This definition of psychology is closer to what we call philosophy of mind and what Kant calls transcendental psychology.
3. Kant also refers to *Rührungen* (stirrings) in other places, which would seem to include both affects and feelings, or perhaps Kant means to imply

that stirrings are merely stirrings and weaker than affects and feelings. It is not clear from his texts. In any case, *Rührungen*, like feelings, as well as other references to internal motion (*die Motion*) are not defined as necessarily bad things, as affects sometimes are.

4. Nevertheless, our translation of *Affect* as affect should not be confused with the purely physiological notion from affect theory. Kant writes: "[C]ertain interior physical feelings are related to the affects, but they are not identical with them since they are only momentary and transitory, leaving no trace behind" (A §79). Affects, on the contrary, leave a mental trace even after their bodily states have ceased. This distinction might also be employed for the difference between emotional and bodily feelings (like pain or hunger). Of course, Kant does not deny that affects and bodily states are related, as we have held all along. It is simply rather the case that bodily sensations do not themselves tell us about the essence of the affect, which can only be understood in terms of a consideration of its causes, whether they be more cognitively explicit or not.
5. Mary Gregor, *Laws of Freedom: A Study of Kant's Method of Applying the Categorical Imperative in the Metaphysik der Sitten* (Oxford: Basil Blackwell, 1963), 73–74.
6. Readers of the *Anthropology*, when looking to the second section, on feeling, will be surprised to see that only emotions are discussed there, not physical feelings like hunger or pain. That strange fact can perhaps be explained by reference to Kant's notion of Anthropology: he takes it to be the study of the universal psychological feature of humans, not a study of their physical capacities.
7. See his *Remark* at A 235.
8. Sherman argues that Kant "did not avail himself of the shared ancient view that emotions are not brute sensations, but states that have evaluative content. This might have made it easier for him to let go of certain rhetoric against the emotions and appreciate even more fully just how reason's project can work through the emotions." Nancy Sherman, *Making a Necessity of Virtue: Aristotle and Kant on Virtue* (New York: Cambridge University Press, 1997), 120. I think that it is only possible to maintain this view of Kant's philosophy of emotion if one limits one's understanding of Kant's theory of emotion to a very limited consideration of his comments about *Affect*—ignoring that they are feelings. If you take his statement that *Affecten* make reflection "difficult or impossible" to mean that they literally make reflection fully impossible, then you have some support for denying him a cognitively wrought theory of emotion. Even still, the case would not be closed since *Affecten* might still have cognitive content from which we are necessarily cut off.
9. For example, Kant discusses the role that imagination plays in making affect contagious, even if only to criticize it. Kant considers the case in which someone, staying up late, becomes emotional about various imagined ideas, only to find that they have faded from memory in the morning: "therefore, the taming of one's imagination, by going to sleep early

in order to rise early, is a very useful rule for the psychological diet" (A 181).

10. Hence in the second section of the *Anthropology* (on the faculty of feeling) affects come off looking normal and not much is recommended to alter them, but in the third section (wherein they are compared to passions) Kant writes that "a mind that is subject to affects and passions is always *ill*" (A 251)!
11. Sorensen discusses three examples of what he calls Kantian "reason-caused" affects: enthusiasm, amazement, and courage. Sorensen writes: "According to the *Metaphysics of Morals,* reason can also produce the following four feelings: conscience, love of human beings, moral feeling and respect" (DV 399–403). The latter two are understood by 1797 as distinct from one another: "moral feeling" is a feeling of approval toward actions required by duty or a feeling of disapproval toward actions prohibited by duty; "respect" is a combination of *both* pleasure and pain directed toward either a person or the law itself. But the third *Critique* contains evidence that Kant thought reason could produce a variety of emotions, seven years before the *Metaphysics of Morals* was published. These new reason-produced emotions include some of the emotions Kant calls *affects*" (115–116). By "reason-caused," Sorensen means virtuous, I believe. He is assuming, we must suppose, that it is common knowledge that Kant holds that there are vicious affects. In isolating these three "reason-caused affects" he is implying that other affects are not caused by reason; he suggests that other affects are caused by desire. Pace Sorensen, desires are not arational, although they might be immoral. We always have some reason for doing/wanting something. See Kelly Sorensen, "Kant's Taxonomy of Emotions," *Kantian Review*, 6 (2002).
12. Sorensen distinguishes between two terms that are both translated as enthusiasm, *die Enthusiasmus* and *die Schwärmerei.* He argues that the latter is negative and the former is positive, for Kant. Sorensen, "Kant's Taxonomy of Emotions," 120.
13. This kind of mistake is regrettable but its possibility is unavoidable. It is by no means evil, which we will greet with the passions. We would not choose to give up or ignore moral affects simply because of this possibility of making mistakes.
14. Sorensen's discussion of courage, as it is discussed in Kant's *Anthropology*, must be taken as an elaboration of Kant's notion of fortitude, which is, of course, his characterization of virtue (Sorensen, "Kant's Taxonomy of Emotions," 121). Much could be said about virtue as the disposition that allows for the quick enactment of moral duty. Being virtuous or vicious refers to the psychologically encompassing notion of character, and so this notion of courage or fortitude as the disposition that enables moral action is a better way to characterize Kantian virtue than as autocracy, or rational self-rule, because the latter notion does not specify the character of the feelings, which should not themselves oppose moral reason. Baxley acknowledges that

"Kantian virtue is not exhausted by the notion of rational self-control over sensibility, for we are also obligated to strengthen and cultivate feelings, desires, and interests in accordance with principles of practical reason." See Anne Margaret Baxley, *Kant's Theory of Virtue* (Oxford: Cambridge University Press, 2010), 5.

15. Letting anger run its course, we might find that it passes through various phases, which might be alarming individually, but which make more sense as a process: rage (the desire for violence), repression of that desire (which is different than the repression of the emotion), a plan for action, attempting to understand the causes of the angering event, reaffirming the goals and values that were thwarted. To suppress these thoughts, then, would be to prevent their rational development.
16. Disgust at another person might also be necessarily vicious, as it is an expression of holding a person, or type of person, in contempt.
17. Guyer concludes, in examining the discussion of freedom and inclination in the "Doctrine of Right," that "there is nothing intrinsically wrong with inclinations, but inclinations are just a part of the ordinary ebb and flow of nature, and there is therefore nothing uniquely valuable about them either." Paul Guyer, *Kant and the Experience of Freedom* (New York: Cambridge University Press), 350.
18. Kant's discussion here is actually rather positive and overlaps with a later section "On the Highest Physical Good," by which he means resting after hard work.
19. For a Kantian account of the duty to stand up for oneself, see Carol Hay, *Kantianism, Liberalism, and Feminism: Resisting Oppression* (New York: Palgrave Macmillan, 2013).
20. Even Gregor's or Ellington's translation of *zähmen* as "subdue" does not necessarily imply repression. That we must calm down so that we can better think through an emotion, so we can decide how to best act on it or coherently express it, does not mean that we must end the emotion, per se, just the overwhelming nature of the affect. It is also perhaps the case that the foregoing discussion about the ebb and flow of emotional experience does imply that in accepting the emotionality of life, one is perhaps not as bothered by it.
21. Kant thinks that shame is more difficult to treat than anger: "It is quite possible to correct a hot temper by inner discipline of the mind; but it is not so easy to devise a plan for overcoming the weakness of a hypersensitive feeling of honor {which manifests itself} in shame" (A 260). It seems true that shame is the manifestation of a character flaw while anger is simply the result of external circumstances. In those cases where anger is the result of a similar character flaw (a hypersensitive feeling of honor) it would be more important, as well as more difficult, to express since expression is likely to hasten critical reflection. For example, if one commonly responds to commonplace criticisms with anger, expressing the thoughts "You are not allowed to criticize me" or "You do not respect me" would help to reveal them as false beliefs.

22. One might object, as Gressis has suggested to me, that anger is itself a sort of threat of violence for the purpose of getting one's own way. If that is the case, it would never be morally acceptable. While we must take context into account (age, gender, position, relationship, etc.), it does not seem to be the case that anger always implies a real threat (violent or otherwise). Violent thoughts might be one component of the process of anger, but for most people those thoughts are not even verbally expressed and they are far from even considered as a real possibility for action. While we navigate anger given the options that are open to us, some of those options are hopefully fair and morally acceptable—mutually agreed upon punishments, for example—while others are not.
23. Many affective theorists define emotions in terms of states of action readiness, such as that being angry makes one likely to yell. This model suggests that emotions have evolved to serve their purposes, but that the emotions of civilized people are often triggered by nonnatural stimuli and are therefore ineffective and inappropriate. Kant's notion of an emotion's purpose is not the same as this idea of action-readiness. The fact that Kant believes that a reasoned response could better serve the purpose, even the purpose of self-preservation, than what we would call an emotional response shows that he believes that emotions cannot be best understood as biological adaptations.
24. Kant also holds that positive affects involve surprise through sensation; if they are also difficult to reflect on, it cannot be because they involve pain. Still, Kant's point might be that the surprise itself, at the mere presence of bodily sensations, perhaps, is disorienting.
25. Baxley, *Kant's Theory of Virtue*, 126–135.
26. Kant writes, "the most thorough and readily available medicine for soothing any pain is the thought, which can well be expected of a reasonable man, that life as such, considered in terms of our enjoyment of it, which depends on fortuitous circumstances, has no intrinsic value at all, and that is has value only in regards to the use to which we put it, the ends to which we direct it. So it is not by fortune but only by wisdom that life can acquire value for us; and its value is, accordingly, within our power. A man who is anxiously concerned about losing his life will never be happy with it" (A 239). Kant here refers to the value of life—not happiness, which depends on the fulfillment of inclination. The thought is that when we have a proper moral orientation in life, minor misfortunes are easier to bear.
27. One should not conclude from this passage that Kant's notion of "the good will" does not relate to his account of moral character or virtue. Having affects is a necessary part of life—that is exactly what Kant is saying here. Also, having good intentions (even the intention to cultivate virtuous emotions) is only one part of virtue; one must also cultivate them. See Baxley's "The Good Will, Moral Worth, and Duty" for a better account of the connection between the notion of the good will, as

it is referred to in the *Groundwork*, and Kant's theory of virtue. Baxley, *Kant's Theory of Virtue*.

28. Kant says this much, and then adds that the Stoic is wrong because we would not want to live without moral affects (A 253–254).
29. Elsewhere Kant asserts that we *always* implicitly judge a pleasure to be good or bad as we experience it (A §64), and so it is unclear to what extent he thinks that we assent to our emotions. That we are both overcome by affect and yet still responsible is a perfectly normal position and poses a question for any consideration of emotion.
30. Nevertheless, they do not immediately cause behaviors; cognitive assent and volition is still necessary. Even pain and pleasure are always mediated by the understanding, for Kant, since they include an evaluation of the pain or pleasure and a consequent submission, but Kant also calls taste and smell the senses of pleasure and we cannot stop ourselves from tasting or smelling certain things; although our understanding might mediate whether or not such tastes or smells are considered to be pleasurable (A§ 21).
31. Sorensen divides up emotions mainly by this criterion. He is right in his argument that emotions, for Kant, can be caused by reasons, regardless of the moral status of this cognition, and because of this it seems that whether or not a feeling leads to a desire for action is not the most important quality of it. See Sorensen, "Kant's Taxonomy of Emotion."
32. Nico Frijda, *The Laws of Emotion* (London: Lawrence Erlbaum, 2007), 15.
33. We see an example of this identification between inclination and immorality in the lectures: "Let thy procedure be such that in all thine actions regularity prevails. What does this restraint imply when applied to the individual? That he should not follow his inclinations." To follow one's inclinations is not only immoral but determined: "he who subjects his person to his inclinations, acts contrary to the ends of humanity, for as a free being he must not be subjected to inclinations, but ought to determine them in the exercise of his freedom" (L 122).
34. A reasonable interpretation here is that, on Kant's account, there are actually two different sources for a feeling of freedom: one rational, one animal. We rationally recognize the pragmatic and moral value of freedom, while, at the same time, we hunger for it instinctively.
35. Affects become passions when they are longer-lived and do not listen to reason. Kant gives the example of a widow who becomes attached to and begins to enjoy mourning—perhaps this affect has come to give her life its only meaning. Kant suggests that such a state might be called an affect, but it is really a passion (A 237).
36. Robert Louden, "Evil Everywhere: The Ordinariness of Kantian Radical Evil," in *Kant's Anatomy of Evil*, ed. Sharon Anderson-Gold and Pablo Muchnik (New York: Cambridge University Press, 2010).

37. If we follow strictly the text of Kant's *Religion within the Boundaries of Mere Reason*, and only this text, we cannot say that this is a conscious choice. See Pablo Muchnik, *Kant's Theory of Evil: An Essay on the Dangers of Self-Love and the Aprioricity of History* (New York: Lexington Books, 2009) for a close reading of this text.
38. This worry is expressed by Claudia Card, *The Atrocity Paradigm: A Theory of Evil* (New York: Oxford University Press, 2002).
39. Scholars of Kant often conclude from the *Religion* essay that Kant holds that people are by nature or inherently evil. This is a mistake. Kant sets up an antinomy between these two opposing positions in the beginning of this work, and like his other treatment of antinomies, he sets it up in order to "solve" it by showing the truth of each position. Unfortunately, his most clear statement charting the middle ground comes in a footnote (to paraphrase a complicated discussion): Ancient philosophers debated about whether or not virtue could be learned and whether or not there is more than one virtue. "To both [questions] they replied with rigorist precision in the negative; and rightly so, for they were considering virtue *in itself*, in the *idea* of reason (how the human being ought to be). [The rigorist position is the position that good and evil are innate, not learned.] If, however, we want to pass moral judgment on this moral being, the human being as he appears, such as experience lets us cognize him, we can then answer both question in the positive. For then he would be judged, not by the scales of pure reason (before a divine court of justice), but according to empirical standards (by a human judge)" (R 25). In other words, if we are asking whether we ourselves or another person are good or evil, we must often conclude that there are ways in which we are good and ways in which we are evil and that we can improve ourselves and become more good. Nevertheless, the essence of evil is choosing against the moral law—even one time.
40. Louden and others are wrong, I believe, to think that there can be no explanation of evil, they say, because the choice to commit an evil act is radically free and thereby not determined by any preceding event. It would be a serious flaw with Kant's notion of freedom if he held that freedom and moral responsibility always coincided with rational inscrutability. Such necessary arbitrariness and irrationality would be an odd tenet for a philosopher who is normally faulted for being too rational. Wood corrects this misinterpretation by likening Kant's phenomenal/noumenal distinction to two ways in which we can see the world (similar to Davidson's two-aspect theory of mind). Similarly, he makes Kant's theory of freedom into compatibility. See Allen Wood, "Kant and the Intelligibility of Evil," in *Kant's Anatomy of Evil*, ed. Sharon Anderson Gold and Pablo Muchnik (New York: Cambridge University Press, 2010); Allen Wood, "Kant's Compatibilism," in *Critical Essays on Kant's Critique of Pure Reason*, ed. Patricia Kitcher (New York: Rowman & Littlefield, 1998). Wood and Anderson-Gold also point to the passions/vices in explaining Kant's moral psychology of evil. See Sharon

Anderson-Gold, "Kant, Radical Evil, and Crimes against Humanity," in *Kant's Anatomy of Evil.* See Neiman for a discussion of whether or not we should try to make evil intelligible, Susan Neiman, *Evil in Modern Thought* (Princeton, NJ: Princeton University Press, 2002).

41. In his *Lectures on Ethics*, Kant taught that "the lessons of morality must be learnt: it ought not to be mixed with solicitations and sensuous incentives; it must be taught apart and free from these; but when the rules of morality in their absolute purity have been firmly grasped, when we have learnt to respect and value them, then, and only then, may such motives be brought into play. They ought not, however, to be adduced as reasons for actions, for they are not moral and the action loses in morality on their account; they ought to serve only as *subsidiara motiva* calculated to overcome the inertia of our nature in the face of purely intellectual conceptions" (L 76).
42. H. Arendt and K. Jaspers, *Correspondence,* 1926–1969 (San Diego: Harcourt, Brace, Jovanovich, 1992), Arendt to Jaspers, March 4, 1951. I am indebted to Louden for this citation. His point here is that this passage, cited by Bernstein, is compatible with Kantian moral theory. See Jay Bernstein, *Radical Evil: A Philosophical Interrogation* (Cambridge, UK: Polity Press, 2002).
43. We are reminded of the theoretical distinction between morality and legality.
44. See Nancy Sherman's *The Untold War* for the suggestion that we do not naturally make these exceptions to the moral law against murder.
45. Sinnott-Armstrong argues that psychopaths lack natural moral feeling and that, for this reason, they cannot be held legally responsible for their actions since they do not actually understand/feel that their actions are wrong. Therefore, he holds that they do not actually make moral judgments and cannot be held morally culpable. He argues that psychopaths should be institutionalized to protect society and given treatment that can perhaps inculcate moral feeling or correct their disability. While it is possible that psychopathology is a congenital neurological disability, it is also possible that the structures of their brains are the result of immoral training and choices. I am grateful to Kelly Sorensen for this reference. Jana Schaich Borg and Walter Sinnott-Armstrong, "Do Psychopaths Make Moral Judgments," in *Handbook on Psychopathology and Law*, ed. Kent Kiehl and Walter Sinnott-Armstrong (New York: Oxford University Press, 2013), 107–128. I agree that whatever "punishment" is most likely to be corrective, while protecting others, is preferable.
46. "[T]he censure of vice...must never break out into complete contempt and denial of any moral worth to a vicious human being; for on this supposition he could never be improved, and this is not consistent with the idea of a human being, who as such (as a moral being) can never lose entirely his predisposition to the good" (DV 463–464).
47. For a Kantian account that highlights the role of history and society for enabling the development of a good character, see Anderson-Gold, "Kant, Radical Evil, and Crimes against Humanity."

7 The Inner Life of Virtue: Moral Commitment, Perfectionism, Self-Scrutiny, Self-Respect, and Self-Esteem

1. Kant might sometimes use religious language, but his moral theory is adamantly not based on religion.
2. Lynne Olson, *Freedom's Daughters: The Unsung Heroines of the Civil Rights Movement from 1830 to 1970* (New York: Scribner Press, 2001).
3. David Halberstan, *The Children* (New York: Fawcett Books, 1999).
4. See Ralph Meerbote, "*Wille* and *Willkür* in Kant's Theory of Action," in *Interpreting Kant*, ed. M. S. Gram (Iowa City: University of Iowa Press, 1982); Henry Allison, *Kant's Theory of Freedom* (New York: Cambridge University Press, 1990); Stephen Engstrom, "Reason, Desire, and the Will," in *Kant's Metaphysics of Morals*, ed. Lara Denis (New York: Cambridge University Press, 2010); Heidi Chamberlin Giannini, "Korsgaard and the *Wille/Willkür* Distinction: Radical Constructivism and the Imputability of Immoral Acts," *Kant Studies Online*, 2013, 72.
5. In Kant's essay "What Is Orientation in Thinking," as well as "What Is Enlightenment?" Kant argues that in order to achieve and sustain political freedom, individuals must govern their own thinking and strive to be objective, not making use of religious or superstitious principles that cannot be shared by all. He argues that thinking for oneself requires that we ask ourselves whether or not it is possible to transform our reason for accepting something into a universal law governing the use of one's reason. Here we see the epistemological side of moral commitment. We might also see this democratic political ideal as another expression of the moral "Kingdom of Ends."
6. Denis explores the different senses in which duties to ourselves have primacy in Kant's philosophy; see Lara Denis, "Freedom, Primacy, and Perfect Duties to Oneself," in *Kant's Metaphysics of Morals*, ed. Lara Denis (New York: Cambridge University Press, 2010).
7. Louden's defense of Kant's moral theory against antimorality and antitheory critics is superior to those suggested here. See Robert B. Louden, *Morality and Moral Theory* (New York: Oxford University Press, 1992).
8. Although the emphasis in Kant's philosophy and within Kant scholarship on principle and consistency may sometimes overshadow the importance of constantly striving toward moral ideals, we must envision practical reason as comfortable with flux and the change of improvement. Indeed, in a world where "ought" only implies "can try," we must sacrifice having our actions match up with our principles in order to ensure that we can have some grasp on correct principles in the first place. Acceptance of our imperfection should not excuse bad behavior but protect the purity of moral ideals.
9. I am swayed by Klein's argument that Aristotle, and not the Stoics, takes virtue to be a stochastic skill; Jakob Klein, "Of Archery and Virtue: Ancient and Modern Conceptions of Value," *Philosopher's Imprint*, 14 (19) (June 2014): 1–16.

10. Supererogationists include J. O. Urmson, David Heyd, and Roderick Chisolm.
11. Hill argues that Kant's notion of imperfect duties takes the place of the category of the supererogatory because it involves the choice of when one is to fulfill them. As Baron suggests, this is not entirely true, because we are always required to take them on as our maxims. In other words, we ought not to choose sometimes to have a virtuous character and sometimes not. See Marcia Baron, *Kantian Ethics Almost without Apology* (Ithaca, NY: Cornell University Press, 1999). Instead Kant writes, "But a wide duty is not to be taken as permission to make exceptions to the maxim of actions but only as permission to limit one maxim of duty by another (e.g., by love of one's neighbor in general by love of one's parents), by which in fact the field for the practice of virtue is widened" (DV 390).
12. See Barbara Herman, *The Practice of Moral Judgment* (Cambridge, MA: Harvard University Press, 1996).
13. R 52 and 171 (of course, Kant has the example of Jesus in his mind in this work). Baron argues that Kant is a rigorist when it comes to perfecting ourselves, that is, that we should always strive to be as good as we possibly can, but that we are allowed more latitude when it comes to helping others. (Baron, *Kantian Ethics Almost without Apology*, especially Chapter 1, "Kantian Ethics and the Supererogatory" and Chapter 2, "Minimal Morality, Moral Excellence, and the Supererogatory.") The duties that we owe to ourselves are more fundamental since a good character is a condition for carrying out the duty to help others, since it facilitates feeling the moral need and judging when and whom to help (Baron, "Latitude in Kant's Imperfect Duties," in *Kantian Ethics Almost without Apology*).
14. In my mind Oskar Schindler is shown to be a truly good person, albeit only by Steven Spielberg's, perhaps fictitious, presentation of his character, when he emotionally regrets not having done more. This feeling is not a sign of neurosis; it is the pain of loss and vulnerability that is a necessary part of any love, in this case, the love of humanity.
15. Susan Wolf, "Moral Saints," in *The Virtues*, ed. Robert B. Kruschwitz and Robert C. Roberts (Boston, MA: Wadsworth, 1987).
16. See Louden, *Morality and Moral Theory*, 49 and Robert Merrihew Adams, The *Virtue of Faith and Other Essays in Philosophical Theology* (New York: Oxford University Press, 1987).
17. Urmson argues that if a moral code is not simple to understand and fulfill, people will give up on being moral entirely; J. O. Urmson, "Saints and Heroes," in *Essays in Moral Theory*, ed. A. Melden (Seattle: University of Washington Press, 1958). Kant, on the contrary, holds that it is not the job of a moral theory to make decisions for people. Kantian moral theory provides guidance for moral reasoning, not the answers. If it is the case that our duties can be prescribed in their detailed specificity without us having to think about them, then it would be possible for

people to only feel the call of duty when they should and will fulfill it. On the other hand, in the real world, where the fulfilling of our imperfect duties requires practical reason, it makes sense that one's feelings of moral compulsion will not always equate with one's decision to act, both because good characters will feel a call to help others before they adjudicate whether or not such is the proper time, and because it is not possible to say when it is objectively morally required that we fulfill our imperfect duties. Kant does not let us off the hook easily: if everyone does not feel that fixing the problems of the world is her responsibility, then there is no chance that moral progress will ever be made. See also Thomas Auxter, *Kant's Moral Teleology* (Macon, GA: Mercer University Press, 1982). I am indebted to Marcia Baron for this reference.

18. Jeanine Grenberg discusses the importance Kant places on "contemplation or attentiveness" at the heart of virtue. Drawing from Engstrom, she identifies this, what I am calling conscientiousness, as the true meaning of the inner strength and freedom that opposes vice. See Jeanine Greenberg, "What Is the Enemy of Virtue?" in *Kant's Metaphysics of Morals*, ed. Lara Denis (New York: Cambridge University Press, 2010) and Stephen Engstrom, "The Inner Freedom of Virtue," in *Kant's Metaphysics of Morals: Interpretive Essays*, ed. Mark Timmons (New York: Oxford University Press, 2002).
19. When discussing gratitude in the *Metaphysics of Morals*, Kant refers to "real self esteem" as "pride in the dignity of humanity of one's own person" (DV 459).
20. In his discussion of lust, Kant argues that "complete abandonment of oneself to animal inclination...deprives him of all respect for himself" (DV 425).
21. For a fuller account of the scope of perfect duties we owe to ourselves as well as varying accounts and justifications of them in Kant's philosophy see Lara Denis, "Freedom, Primacy, and Perfect Duties to Oneself," In *Kant's Metaphysics of Morals*, ed. Lara Denis (New York: Cambridge University Press, 2010). Her *Moral Self-Regard* also has a discussion of both duties we owe to our natural being and duties we owe to our moral being; Lara Denis, *Moral Self-Regard: Duties to Oneself in Kant's Moral Theory* (New York: Routledge, 2001).
22. We should note that Kant believes in the death penalty because he holds that this punishment best respects the humanity and rationality of the criminal, not because he takes the criminal to be contemptible. We might disagree about whether or not attempts at criminal reform can better demonstrate respect for humanity, but the point is that the punishment is addressed to the rational nature of the criminal and his recognition of his desert, not his past deeds. For a fuller treatment of this topic see Wood's discussion of Kant's retributivist position in Allen Wood, "Punishment, Retribution, and the Coercive Enforcement of Right," in *Kant's Metaphysics of Morals*, ed. Lara Denis (New York: Cambridge University Press, 2010).

23. Kant is also worried about the lack of self-respect that might follow from lowering oneself in comparison with a notion of God. Kant refers to prostrating oneself before God as contrary to this duty to oneself; he also suggests that false belief in God may be the ground of spurious moral principles (DV 436 and DV 430). While, of course, a human being is not a god, and one's powers are severely limited, humanity itself deserves respect.
24. Robert Gressis discusses two false notions of self-esteem that lead to the justification of evil: the idea that one is as good as or better than one's peers, and the idea that one deserves a break from moral rules sometimes. Robert Gressis, "How to Be Evil: The Moral Psychology of Immorality," in *The New Kant*, ed. Pablo Muchnik (Cambridge, UK: Cambridge Scholars, 2008).
25. Kant seems to be self-contradictory on this point, arguing first that self-esteem is a member of the class of natural feelings that we cannot be said to have a duty to acquire (DV 399) and then that "self-esteem is a duty of man to himself" (DV 435). Furthermore, if we did not naturally respect the moral law, we would have no basis for self-respect (DV 402). The solution must involve delineating different senses of the term, as I have here.
26. To morally evaluate oneself in this way is a duty, related to self-scrutiny: "Impartiality in appraising oneself in comparison with the law, and sincerity in acknowledging to oneself one's inner moral worth or lack of worth are duties that follow directly from this first command to cognize oneself" (DV 441–442).
27. Drawing an analogy with the Stoics, Beck calls this loftiness "equanimity"; Lewis White Beck, *A Commentary on Kant's Critique of Practical Reason* (Chicago: University of Chicago Press, 1960) 230.
28. Ibid., 229.
29. Kant then discusses sweet merit and bitter merit: the former is the product of an entirely successful moral win, as it were, and the latter is the product of a moral deed that is accompanied by some other unfortunate consequence, like ingratitude or personal loss, for example (DV 391). In the second case, we perhaps deserve to feel more satisfied with ourselves since we have undergone suffering in order to do what is right, but might simply feel regret at the unfortunate circumstances. Nevertheless, we at least become aware of our own strength and goodness.
30. Kant rediscovers this unity in the third *Critique* under the name of aesthetic feeling and he, perhaps mistakenly, originally theorized it in the first *Critique*, by making the categories of the Understanding dependent on the forms of intuition.
31. In the *Anthropology*, Kant describes resting after work as the "highest physical good" (§87).
32. Rather than, or perhaps in addition to, facilitating activities at which children can succeed, then, it would be more conducive to positive self-esteem to create situations in which they can exercise good qualities, like perseverance and generosity.

33. See CPrR 161. The second *Critique* is, in general, more focused on the dichotomy between inclination and duty, and so it casts self-esteem in terms of the special status that humans achieve through being aware of their ability to transcend inclinations.
34. CJ §83 shines some particularly helpful light on Kant's meaning of the word happiness: happiness is a "mere idea" to which we attempt to make ourselves "adequate under merely empirical conditions (which is impossible)." It is a deficient idea that is necessarily tied to the shortsightedness of passion/inclination and those things to which one is naturally/automatically directed. Even the term "true happiness" for Kant still involves the exclusion of moral reason. Why does Kant insist on defining happiness derogatorily? It seems as though it is merely a polemical device aimed at sharpening our attention to the sublimity of morality, as well as separating himself from other moral theories he deems flawed.
35. It is this cruel fact about the world that, according to him, makes heaven an innate idea of morality. There must be a place, we think, in which goodness is duly rewarded. Kant writes: "The highest good of a possible world" consists in "happiness distributed quite exactly in proportion to morality" (CprR 110); it perhaps goes without saying that we do not live in such a world.
36. Morally worthy happiness may be a strange idea. What value is left for the fulfillment of inclination if one recognizes that the inclinations must be subordinate to morality? It seems that morally worthy happiness must assume some higher definition of happiness, not just the synthetic idea of happiness conditioned on prior moral goodness. Kant does consider the ways in which virtue is its own reward, and hence its own brand of happiness, as we shall see shortly.
37. It may seem strange that someone can deserve or not deserve to be happy; Kant more properly means that if one happens to be happy, then one can feel good about it. Nevertheless, there is little difference between this and the idea that someone does not deserve his or her wealth, good fortune, or good health. It may seem strange, but it is also empowering because it challenges our culture's perverse pursuit of wealth and health. It also explains the commonplace experience of having everything one wants and yet not feeling "happy"—self-satisfaction, in the Kantian sense.
38. See CPrR 57–67.
39. Ibid.
40. Most notably by Jürgen Habermas, *Moral Consciousness and Communicative Action*, trans. Christian Lenhardt and Shierry Weber Nicholsen (Cambridge: MIT Press, 1995), 172.
41. I do not believe that Stevenson's theory of emotivism is recognizable as emotivism; see C.L. Stevenson, *Ethics and Language* (New Haven, CT: Yale University Press, 1944). In Stevenson's own words, his theory is an "analytic or methodological study" of normative ethics and not itself normative (p. 1.) Although emotivism is also clearly a metaethical theory,

it is also normative, as it implies that moral argumentation is irrational insofar as it entails attempting to change people's values, not just their beliefs about facts. Stevenson's emotivism has no such implication, as he agrees with the Rachels's account of moral reasoning (see, e.g., 27, 36, 139; see 173 for a discussion of the logical-like validity of ethical judgments). Stevenson's emotivism is more of an empirical study of ethical discourse and does not tell us anything that runs counter to common sense. Nor is it particularly interested in emotion as a key concept. Still, Stevenson argues that his own theory has much in common with Ayer's, except that it does not intend to disparage ethical argument and inquiry as unscientific. That which it shares with Ayer falls to the same criticisms as Ayer does, and his differences, as he strays from noncognitivism, make him unrecognizable as an emotivist. In addition, Stevenson seems to be even less equipped to maintain a distinction between the expression, or what he calls "giving vent" to (37–38), and assertion of emotion. It is unclear whether he intends to uphold Ayer's distinction, since he translates ethical statements into statements about subjective approval. One might argue that his distinction between descriptive and emotive meaning, or the disposition of words to affect cognition and the disposition of words to affect feeling (71), equaling roughly a distinction between denotation and connotation, does this job. Although, for Ayer, the assertion of emotion counts as a factual statement, and therefore has descriptive meaning in Stevenson's sense, Stevenson's notion of emotive meaning is a property of the sign, not of the speaker, making it more like a connotation, and is therefore unrelated to Ayer's distinction.

42. In fact, some people think that morality is defined as "our personal values," and there is a prevalent intellectual myth that "ethics" is distinguished from "morality" because the first is based in a society and the second is based in an individual. Nevertheless, even our personal attempts to improve ourselves would be derailed by our lack of understanding the nature of morality since improving myself would require that I believe that something else actually *is* better. If something were only good *for me*, then I could easily change my way of being or thinking to avoid needing to change. To set the record straight: "ethics" and "morality" are synonyms; morality is the Latinate. Ethics comes from the Greek *ethos*, but it is not the same as the English word "ethos," which refers to cultural practices, nor were the most important Greek moralists cultural relativists.

43. In fact, it is *morally disrespectful* to think that some other types of people are so different that they do not experience the demands of reason and morality.

44. Calhoun discusses possible conflicts between intellectual and experiential beliefs in her "Cognitive Emotions?" My discussion here about knowing oneself and acting consistently parallels her discussion about working toward aligning experiential beliefs with intellectual beliefs, thereby overcoming emotion/belief conflicts.

45. This example sheds light on the interpretive debate concerning Kant's examples of acting out of duty as opposed to acting in conformity with duty. Many who defend Kant argue that, with his example in the *Groundwork* of the person continuing his life against all inclination, Kant only means to say that it is in the cases where duty and inclination conflict that we can *be sure* that we are acting from duty. When they are not in conflict, we might still be acting from duty, it is just difficult to tell. My example suggests that it is likely the case that when inclination and duty coincide we are acting out of duty less often than we would like to think.
46. Of course, it is possible that I have merely held this moral ideal unthinkingly. In that case, I should test my belief by evaluating the reasons that I have for it. Even if my principles are wrong, acting on them helps me to evaluate them. Perhaps, in perpetually offering rides to people, I begin to feel taken advantage of. I therefore begin to doubt whether or not it is good to be charitable in this way *all the time*, and perhaps I decide to qualify, and thereby alter, the principle.
47. Henry Allison, *Kant's Theory of Freedom* (Cambridge, UK: Cambridge University Press, 1990), 40.
48. Roger Sullivan, *Introduction to Kant's Metaphysics of Morals* (Cambridge, UK: Cambridge University Press, 1996).
49. Jeanine Grenberg, "What Is the Enemy of Virtue?"
50. J. David Velleman, *Practical Reflection* (New York: Oxford University Press, 1989), 306.
51. Alan Gewirth, *Self-Fulfillment* (Princeton, NJ: Princeton University Press, 1998), 74.
52. Harry Frankfurt, "Freedom of the Will and the Concept of a Person," in *The Importance of What We Care About* (Cambridge, UK: Cambridge University Press, 1987).
53. Susan Wolf, *Freedom within Reason* (New York: Oxford University Press, 1990).
54. Frankfurt, *The Importance of What We Care About*, 86.
55. Herman, *The Practice of Moral Judgment*, 197–198. See also Diane Williamson, "Familial Duties and Emotional Intelligence" in *The Ethics of the Family*, ed. Stephen Scales, Adam Potthast, and Linda Oravecz (Cambridge, UK: Cambridge Scholars, 2010).]
56. Robert Solomon, "Emotions and Choice," in *Explaining Emotions*, ed. Amélie Oksenberg Rorty (Oakland: University of California Press, 1980).
57. Felicitas G. Munzel, *Kant's Conception of Moral Character* (Chicago: University of Chicago Press, 1999), 305.
58. Nevertheless, Kant does try to justify the possible purity of practical reason by arguing that the solution to the third Antinomy relies on it. See Beck, *A Commentary on Kant's Critique of Practical Reason*, 170–175.
59. O'Neill argues that all reason, speculative and practical, is subordinate to moral reasoning. Onora O'Neill, *Constructions of Reason: Explorations*

of Kant's Practical Philosophy (Cambridge, UK: Cambridge University Press, 1989).

8 A Morally Informed Theory of Emotional Intelligence

1. Carolyn MacCann et al., "The Assessment of Emotional Intelligence: On Frameworks, Fissures, and the Future," in *Measuring Emotional Intelligence: Common Ground and Controversy*, ed. Glenn Geher (Hauppauge, NY: Nova Science, 2004), 26. The MSCEIT (Mayor Salovey Caruso Emotional Intelligence Test) and the EQ-i (Emotional Quotient Inventory) were shown to have a correlation of 0.20 in one study. Marc A. Brackett and John D. Mayer, "Convergent, Discriminant, and Incremental Validity of Competing Measures of Emotional Intelligence," *Personality and Social Psychology Bulletin*, 29 (9) (2003): 1147–1158.
2. Peter Salovey and John D. Mayer, "Emotional Intelligence," *Imagination, Cognition and Personality*, 9 (3) (1989–1990): 189.
3. Edward L. Thorndike and S. Stein, "An Evaluation of the Attempts to Measure Social Intelligence," *Psychological Bulletin* 34 (1937): 275–285.
4. Lisa F. Barrett and Peter Salovey, "Introduction," in *The Wisdom in Feeling: Psychological Processes in Emotional Intelligence*, ed. Lisa F. Barrett and Peter Salovey (New York: Guilford Press, 2002), 2.
5. Reuven Bar-On, *Bar-On Emotional Quotient Inventory: Technical Manual* (North York, ON: Multi-Health System, 1997).
6. Reuben Bar-On and J. Parker, eds., *The Handbook of Emotional Intelligence: Theory, Development, Assessment, and Application at Home, School and in the Workplace* (San Francisco: Jossey-Bass, 2000), 365.
7. R. E. Boyatzis, D. Goleman, and K. Rhee, "Clustering Competence in Emotional Intelligence: Insights from the Emotional Competence Inventory," in Bar-On and Parker, *The Handbook of Emotional Intelligence*.
8. Daniel Goleman, *Emotional Intelligence: Why It Can Matter More Than IQ* (New York: Bantam Books, 1995), xii and 34.
9. Ibid., 36–37.
10. Ibid., xii.
11. Ibid., 14.
12. Kevin Murphy and L. Sideman, "What Is This Thing Called Emotional Intelligence?" in *A Critique of Emotional Intelligence: What Are the Problems and How Can They Be Fixed?*, ed. Kevin Murphy (New York: Psychology Press, 2006), 63.
13. Peter Salovey and D. Sluyter, eds., *Emotional Development and Emotional Intelligence: Educational Implications* (New York: Basic Books, 1997).
14. Murphy, *A Critique*, xii.

15. M. Zeidner and G. Matthews, "Personality and Intelligence," in *Handbook of Human Intelligence*, ed. R. J. Sternberg (Cambridge, UK: Cambridge University Press, 2000).
16. Jeffrey Conte and Michelle Dean, "Can Emotional Intelligence Be Measured?" in Murphy, *A Critique.*
17. Introversion, as long as it is not debilitating shyness, does not seem to be a bad thing. There is less reason to think that measures of emotional intelligence should be biased toward extroversion.
18. I worry that some education reformers place too much importance on the notion of character, thereby implying that it is relatively inalterable. See Paul Tough, *How Children Succeed: Grit, Curiosity, and the Hidden Power of Character* (Boston, MA: Houghton Mifflin Harcourt, 2012).
19. C. Spearman, "'General Intelligence,' Objectively Determined and Measured," *American Journal of Psychology* 15 (1904): 201–293; A. Binet and T. Simon, "Methodes nouvelles pour le diagnostic du niveau intellectual des anormaux," *L'Annee Psychologique* 11 (1905): 191–244. Brody argues that heritability is wrongly confused with immutability, where long-term population wide changes, known as the Flynn-effect, show that intelligence is malleable in principle. James Flynn, "IQ Gains Over Time: Towards Finding the Causes," in *The Rising Curve: Long-Term Gains in IQ and Related Measures*, ed. U. Neisser (Washington, DC: American Psychological Association, 1998), 25–66; Nathan Brody, "Beyond g," in Murphy, *A Critique of Emotional Intelligence.*
20. R. Plomin et al., *Behavioral Genetics,* 4th ed. (New York: Freeman, 2001). General intelligence has become an infamous concept, as it is notoriously accused of reductionism, determinism, racism, classism, and conservatism. Stephen J. Gould's popular *The Mismeasure of Man* (New York: W. W. Norton, 1996) makes a number of these arguments.
21. Brody draws this conclusion from research conducted by M. J. Schulte, M. J. Ree, and T. R. Caretta, "Emotional Intelligence: Not Much More Than G and Personality," *Personality and Individual Differences* 37 (2004): 1059–1068.
22. J. D. Mayer and P. Salovey, "What Is Emotional Intelligence?" in Salovey and Sluyter, *Emotional Development and Emotional Intelligence*, 6.
23. Gardner posits a variety of different types of intelligence—at least eight kinds: logical, linguistic, spatial, musical, kinesthetic, naturalistic, interpersonal, intrapersonal—rather than just the logical and linguistic intelligence that is represented by IQ. Howard Gardner, *Intelligence Reframed: Multiple Intelligences for the 21st Century* (New York: Basic Books, 1999). Here we can see that the vagueness and broadness of Goleman's notion is actually its selling point, like a horoscope, giving people more ways in which they can identify with the definition. The more kinds of intelligence there are, the more people get to count as intelligent. See Thomas Armstrong, *Awakening Your Child's Natural Genius* (New York: Putnam, 1991).

24. Unfortunately, not only is this problematic notion of intelligence thought to constrain potential, discouraging educative effort, it ironically also dampens the effort spent on those thought to be intelligent. Although it is intended to secure them more attention, since intelligent students are more capable of learning on their own (even by Goleman's definition), they would require less attention and one-on-one instruction. In reality, the idea of intelligence has no place in the classroom, where we must assume that all students can be and need to be taught. See also Harold Stevenson and James Stigler, *Learning Gap: Why Our Schools Are Failing and What We Can Learn from Japanese and Chinese Education* (New York: Touchstone, 1992). For a contrary argument, see L. Cronbach and R. Snow, *Aptitudes and Instructional Methods: A Handbook for Research on Interactions* (New York: Irvington, 1977).
25. Goleman identifies emotional intelligence with the ability to delay gratification and discusses an experiment designed to show that this ability is constant from age four on to late adolescence. This experiment tests four-year-olds by asking them if they would rather have one marshmallow at the present time or wait 15–20 minutes for two marshmallows. Reported in Yuichi Schoda, Walter Mischel, and Philip K. Peake, "Predicting Adolescent Cognitive and Self-Regulatory Competencies from Preschool Delay Gratification," *Developmental Psychology*, 26 (6) (1990): 978–986. Those who choose to wait are labeled as possessing the ability to delay gratification. The researchers then correlated this ability with later traits, such as ability to cope with stress, concluding that those who exhibited the ability to delay gratification were better off later in life. Goleman uncritically accepts the methodology and conclusions of this study, writing: "Which of these choices a child makes is a telling test; it offers a quick reading not just of character, but of the trajectory that child will probably take through life" (81). Goleman goes further and suggests that this test illustrates the essence of emotional intelligence, which is self-control. Nowhere does Goleman suggest that instruction in the delay of gratification would be possible. Furthermore, this test does not control for diet, which is problematic: a child who does not know what a marshmallow is would both have less trouble resisting one and better long-term health and behavior.
26. Goleman, *Emotional Intelligence*, 82; citing Jack Block's manuscript from research conducted at UC Berkeley.
27. Brody, "Beyond g."
28. D. R. Caruso, B. Bienn, and S.A. Kornacki, "Emotional Intelligence in the Workplace," in *Emotional Intelligence in Everyday Life*, 2nd ed., ed. J. Ciarrochi, J. Forgas and J.D. Mayer (Philadelphia, PA: Psychology Press, 2006), 202; Murphy and Sideman, "What Is This Thing Called Emotional Intelligence?" 39.
29. Murphy and Sideman, "What Is This Thing Called Emotional Intelligence?" 27.

30. Ibid., 28.
31. Brody, "Beyond g," 178–179.
32. J. Ciarrochi et al., "Improving Emotional Intelligence: A Guide to Mindfulness-Based Emotional Intelligence Training," in *Applying Emotional Intelligence*, ed. J. Ciarrochi and J. Mayer (New York: Psychology Press, 2007).
33. Ibid., 89.
34. Marc A. Brackett and N. Katulak, "Emotional Intelligence in the Classroom: Skill-Based Training for Teachers and Students," in Ciarrochi et al., *Applying Emotional Intelligence*,.
35. Overall, in this book I have argued that a good emotional comportment requires a correct latent theory of emotion. I have also highlighted the role that meta-judgments plays in emotional experience. It is likely that a person who is emotionally fluent is free from incorrect and complicating theories and meta-judgments.
36. M. Buckley and C. Saarni, "Skills of Emotional Competence: Pathways of Development," in Ciarrochi et al., *Emotional Intelligence in Everyday Life*, 52.
37. Ibid., 55–56.
38. Fitness notes that it only takes one emotionally intelligence partner to bring emotional intelligence to a marriage. This supports the conclusion that emotional intelligence is learnable. J. Fitness, "The Emotionally Intelligent Marriage," in Ciarrochi et al., *Emotional Intelligence in Everyday Life*.
39. Furthermore, one problem with emotional intelligence tests is the need to have a good theory of emotion in order to be able to write the test and know what counts as a correct answer. When you ask people to "solve" an emotion problem, for example, you need to know beforehand what the right solution is in order to know whether or not they solved it correctly. See Zeidner et al., "Measure for Emotional Intelligence Measures," in *What We Know about Emotional Intelligence: How It Affects Learning, Work, Relationships, and Our Mental Health*, ed. M. Zeidner, G. Matthews, and R. Roberts (Cambridge: MIT Press, 2009).
40. See Matthews et al., "What Is This Thing Called Emotional Intelligence," for a review of the literature. There is some evidence that Goleman's notion of emotional intelligence, as it is measured by a short version of the MEIS, correlates with customer service skills, although perhaps at the expense of productivity. See also A. E. Feyerherm and C. L. Rice, "Emotional Intelligence and Team Performance: The Good, the Bad, and the Ugly," *International Journal of Organizational Analysis* 10 (2002): 343–362.
41. For the correlation between EI and faking positive emotions see: T. Cage, C. S. Daus, and K. Saul, "An Examination of Emotional Skill, Job Satisfaction, and Retail Performance," presented at the 19th Annual Society for Industrial/Organizational Psychology (Washington, DC, 2005); discussed in C. S. Daus, "The Case for the Ability Based

Model of Emotional Intelligence," in Murphy, *A Critique of Emotional Intelligence.*

42. See A. R. Hochschild, *The Managed Heart: Commercialization of Human Feeling* (Berkeley: University of California Press, 1983). See Daus, "The Case for the Ability Based Model of Emotional Intelligence" for a discussion of emotional labor and emotional intelligence.
43. Alison Jagger, "Love and Knowledge: Emotion in Feminist Epistemology," in *Gender, Body, Knowledge*, ed. Alison Jagger and Susan R. Bordo (New Jersey: Rutgers University Press, 1989).
44. Child abuse might be an example of an "emotional hijacking," and such an example would reveal the perversity of Goleman's idea that we need to "control" our emotions. A parent should never need to "control" the desire to abuse his child; a parent should never experience such a desire at all.
45. Baruch Spinoza, *Ethics*, trans. Samuel Shirley, ed. Seymour Feldman (Indianapolis, IN: Hackett, 1992), 260.
46. Ibid., 128.
47. Salovey and Mayer, "What Is Emotional Intelligence?" 4–5.
48. Ciarrochi et al., "Improving Emotional Intelligence."
49. Goleman tellingly can find no flaw with repression: *Emotional Intelligence*, 75–77.
50. We can take the argument even further and argue that, in some situations, it is morally unacceptable to ignore emotional communication and information. Imagine a situation, for example, where we would normally believe that consent is morally required; clearly, accepting merely verbal assent when there is emotional information to the contrary would be morally unacceptable. This might seem like a special situation, but it is necessary for us to take a step back and realize that all of our interactions and relationships are emotional, even those between relative strangers. The empathy referred to by the notion of emotional intelligence demands that we be aware of this unspoken dimension: it does not take a mind reader to know what someone else might be thinking and feeling, and yet we are often afraid to address it. We are morally called on to accept and validate the reality of the unspoken, but, again, emotions are not just the unspoken. We are also morally called on to act with courage, and stand by our convictions, in our explicit, verbal exchanges. Without an understanding of the moral importance of emotional intelligence, morality comes to refer to an abstract and impersonal domain, if it does indeed leave us with any duties at all.
51. See Ciarrochi et al., "Improving Emotional Intelligence."
52. Zeidner et al., *What We Know About Emotional Intelligence.*
53. See chapter 2.
54. Furthermore, it is likely that researchers feel compelled to maintain that emotional intelligence is corrigible because we feel loath to excuse a person's emotional failure and instead maintain a sense of moral culpability.
55. K. Kristjánsson, "'Emotional Intelligence' in the Classroom? An Aristotelian Technique," *Education Theory* 56 (2006): 39–56; L.

Waterhouse, "Inadequate Evidence for Multiple Intelligences, Mozart Effect and Emotional Intelligence Theories," *Educational Psychologist* 41 (2006): 247–255. I am indebted to Zeidner et al. for these references.

56. R. E. Boyatzis and F. Sala, "Assessing Emotional Intelligence Competencies," in *The Measurement of Emotional Intelligence*, ed. Glenn Geher (Hauppauge, NY: Novas Science, 2004).
57. J. Fitness, "The Emotionally Intelligent Marriage," in Ciarrochi et al., *Emotional Intelligence in Everyday Life*, 132.
58. C. Saarni, "Emotional Competence and Self-Regulation in Childhood," in Salovey and Sluyter, *Emotional Development and Emotional Intelligence*,, 35 and 39.
59. Ibid., 38.
60. Ibid., 39.
61. See also Marc A. Brackett, R. M. Warner, and J. S. Bosca, "Emotional Intelligence and Relationship Quality among Couples," *Personal Relationships* 12 (2005): 197–212; and P. N. Lopes et al., "Emotional Intelligence and Social Interaction," *Personality and Social Psychology Bulletin* 30 (2004): 1018–1034.
62. M. M Tugade and B. L. Fredrickson, "Positive Emotions and Emotional Intelligence," in Barrett and Salovey, *The Wisdom in Feeling*, 333. They draw from H. Tennen and G. Affleck, "Finding Benefits in Adversity," in *Coping: The Psychology of What Works*, ed. C. R. Snyder (New York: Oxford University Press,1999) and Ronnie Janoff-Bulman, *Shattered Assumptions: Towards a New Psychology of Trauma* (New York: Free Press, 1992).
63. John Mayer, "A New Field Guide to Emotional Intelligence," in Ciarrochi et al., *Emotional Intelligence in Everyday Life*, 24. Salovey and Mayer acknowledge that conceiving of and measuring emotional intelligence requires that we have some "right answers" about emotional reactions. They choose to emphasize the extent to which these answers are not possible because of cultural relativity. Mayer and Salovey, What Is Emotional Intelligence?" 9.
64. Fitness, "The Emotionally Intelligent Marriage."

CONCLUSION

1. An Aristotelian example of a statue: the marble is the material cause, the sculptor is the efficient cause, the image is the formal cause, and the purpose of worshipping the god is the final cause.
2. If such were the case, it is more likely that the neurobiologist would have reason to second-guess his findings than the philosopher since the former is dependent on first-person accounts of experience in order to postulate the meaning of brain scans in the first place.
3. Jesse Prinz, *The Emotional Construction of Morals* (New York: Oxford University Press, 2011), 1.
4. John Rawls, *A Theory of Justice* (Cambridge, MA: Belknap Press, 1971).

Bibliography

Adams, Robert Merrihew. The *Virtue of Faith and Other Essays in Philosophical Theology*. New York: Oxford University Press, 1987.

Adorno, Theodor and Max Horkheimer. *Dialectic of Enlightenment*. New York: Continuum, 1976.

Adorno, Theodor. *Aesthetic Theory*. Minneapolis: University of Minnesota Press, 1998.

Aknin, L., M. Norton, and E. Dunn. "From Wealth to Well-Being? Money Matters, but Less Than People Think." In *Journal of Positive Psychology*, 4 (6) (2009): 523–527.

Allison, Henry. *Kant's Groundwork for the Metaphysics of Morals: A Commentary*. New York: Oxford University Press, 2011.

Allison, Henry. *Kant's Theory of Freedom*. New York: Cambridge University Press, 1990.

Allison, Henry. *Kant's Theory of Taste, A Reading of the Critique of Aesthetic Judgment*. New York: Cambridge University Press, 2001.

Anderson-Gold, Sharon. "Kant, Radical Evil, and Crimes against Humanity." In *Kant's Anatomy of Evil*, edited by Pablo Muchnik and Sharon Anderson-Gold. New York: Cambridge University Press, 2010.

Annas, Julia. *The Morality of Happiness*. New York: Oxford University Press, 1993.

Arendt, Hannah. *Lectures on Kant's Political Philosophy*. Chicago: University of Chicago Press, 1989.

Arendt, Hannah and K. Jaspers. *Correspondence* 1926–1969. San Diego: Harcourt, Brace, Jovanovich, 1992.

Armstrong, Thomas. *Awakening Your Child's Natural Genius*. New York: Putnam, 1991.

Auxter, Thomas. *Kant's Moral Teleology*. Macon, GA: Mercer University Press, 1982.

Averill, James. "Emotion and Anxiety: Sociocultural, Biological, and Psychological Determinants." In *Explaining Emotions*, edited by Amélie Rorty. Oakland: University of California Press, 1980.

Bar-On, Reuben. *Bar-On Emotional Quotient Inventory: Technical Manual*. North York: Multi-Health System, 1997.

Bar-On, Reuben and J. Parker, eds. *The Handbook of Emotional Intelligence: Theory, Development, Assessment, and Application at Home, School and in the Workplace*. San Francisco: Jossey-Bass, 2000.

Baron, Marcia. *Kantian Ethics Almost without Apology*. Ithaca, NY: Cornell University Press, 1999.

Baron, Marcia. "Love and Respect in the Doctrine of Virtue." In *Kant's Metaphysics of Morals*, edited by Mark Timmons. New York: Oxford University Press, 2002.

Barrett, Lisa F. and Peter Salovey. "Introduction." In *The Wisdom in Feeling: Psychological Processes in Emotional Intelligence*, edited by Lisa F. Barrett and Peter Salovey. New York: Guilford Press, 2002.

Baxley, Anne Margaret. *Kant's Theory of Virtue*. New York: Cambridge University Press, 2010.

Bechara, Antoine, Hanna Damasio, Daniel Tranel, and Antonio Damsio. "Deciding Advantageously before Knowing the Advantageous Strategy." *Science*, 275 (1997): 1293–1294.

Beck, Lewis White. *A Commentary on Kant's Critique of Practical Reason*. Carbondale: University of Chicago Press, 1960.

Beiser, Frederick. "Moral Faith and the Highest Good." In *The Cambridge Companion to Kant and Modern Philosophy*, edited by Paul Guyer. New York: Cambridge University Press, 2006.

Berlin, Isaiah. *Liberty*. Oxford: Oxford University Press, 2002.

Bernstein, Jay. *Radical Evil: A Philosophical Interrogation*. Cambridge: Polity Press, 2002.

Betzler, Monika. "Kant's Ethics of Virtue: An Introduction." In *Kant's Ethics of Virtue*, edited by Monika Betzler. Boston, MA: Walter de Gruyter, 2008.

Binet, A. and T. Simon. "Methodes nouvelles pour le diagnostic du niveau intellectual des anormaux." *L'Annee Psychologique*, 11 (1905): 191–244.

Blum, Lawrence. *Friendship, Altruism, and Morality*. New York: Routledge & Kegan Paul, 1980.

Borg, Jana Schaich and Walter Sinnott-Armstrong, "Do Psychopaths Make Moral Judgments?" In *Handbook on Psychopathology and Law*, edited by Kent Kiehl and Walter Sinnott-Armstrong. New York: Oxford University Press, 2013.

Borges, Maria. "Physiology and the Controlling of Affects in Kant's Philosophy." *Kantian Review*, 13 (2) (2008): 46–66.

Borges, Maria. "What Can Kant Teach Us about Emotions?" *Journal of Philosophy*, 101 (3) (2004): 140–158.

Bowen, Murray. *Family Therapy in Clinical Practice*. New York: Jason Aronson, 1994.

Boyatzis, R. E., D. Goleman, and K. Rhee. "Clustering Competence in Emotional Intelligence: Insights from the Emotional Competence Inventory." In *The Handbook of Emotional Intelligence: Theory, Development, Assessment, and Application at Home, School and in the Workplace*, edited by Reuben Bar-On and J. Parker. San Francisco: Jossey-Bass, 2000.

Boyatzis, R. E. and F. Sala. "Assessing Emotional Intelligence Competencies." In *The Measurement of Emotional Intelligence*, edited by Glenn Geher. Hauppauge, NY: Novas Science, 2004.

Brackett, Marc A. and John D. Mayer, "Convergent, Discriminant, and Incremental Validity of Competing Measures of Emotional Intelligence." *Personality and Social Psychology Bulletin*, 29 (9) (2003): 1147–1158.

Brackett, Marc A., R. M. Warner, and J. S. Bosca. "Emotional Intelligence and Relationship Quality among Couples." *Personal Relationships*, 12 (2005): 197–212.

Brackett, Marc A. and N. Katulak. "Emotional Intelligence in the Classroom: Skill-Based Training for Teachers and Students." In *Applying Emotional Intelligence*, edited by J. Ciarrochi and J. Mayer. New York: Psychology Press, 2007.

Brody, Nathan. "Beyond g." In *A Critique of Emotional Intelligence: What Are the Problems and How Can They Be Fixed?* edited by Kevin Murphy. New York: Psychology Press, 2006.

Buckley, M. and C. Saarni. "Skills of Emotional Competence: Pathways of Development." In *Emotional Intelligence in Everyday Life*, 2nd edition, edited by J. Ciarrochi, J. Forgas, and J. Mayer. New York: Psychology Press, 2006.

Cage, T., C. S. Daus, and K. Saul. "An Examination of Emotional Skill, Job Satisfaction, and Retail Performance." Paper presented at the 19th Annual Society for Industrial/Organizational Psychology, Washington, DC, 2005.

Calhoun, Cheshire. "Cognitive Emotions?" In *What Is an Emotion?* edited by Robert Solomon and Cheshire Calhoun. New York: Oxford University Press, 1984.

Calhoun, Cheshire and Robert Solomon. "Introduction." In *What Is an Emotion?* edited by Cheshire Calhoun and Robert Solomon. New York: Oxford University Press, 1984.

Cannon, W. B. "The James-Lange Theory of Emotion: A Critical Examination and an Alternative Theory." In *The Nature of Emotion*, edited by M. B. Arnold. New York: Penguin, 1968.

Card, Claudia. *The Atrocity Paradigm: A Theory of Evil.* New York: Oxford University Press, 2002.

Carey, Benedict. "The Muddled Tracks of All Those Tears." *The New York Times*, February 2, 2009.

Caruso, D. R., B. Bienn, and S. A. Kornacki. "Emotional Intelligence in the Workplace." In *Emotional Intelligence in Everyday Life*, 2nd edition, edited by J. Ciarrochi, J. Forgas, and J. D. Mayer. Philadelphia, PA: Psychology Press, 2006.

Carruthers, Mary. *The Craft of Thought: Meditation, Rhetoric, and the Making of Images, 400–1200.* New York: Cambridge University Press, 1998.

Churchland, Patricia. *Braintrust: What Neuroscience Tells Us about Morality.* Princeton, NJ: Princeton University Press, 2011.

Ciarrochi, Joseph and John T. Blackledge. "Improving Emotional Intelligence: A Guide to Mindfulness-Based Emotional Intelligence Training." In *Applying Emotional Intelligence*, edited by J. Ciarrochi and J. Mayer. New York: Psychology Press, 2007.

Clewis, R. R. *The Kantian Sublime and the Revelation of Freedom*. New York: Cambridge University Press, 2009.

Conte, Jeffrey and Michelle Dean. "Can Emotional Intelligence Be Measured?" In *A Critique of Emotional Intelligence: What Are the Problems and How Can They Be Fixed?* edited by Kevin Murphy. New York: Psychology Press, 2006.

Cooper, John M. *Reason and Emotion: Essays on Ancient Moral Psychology and Ethical Theory*. Princeton, NJ: Princeton University Press, 1999.

Cosmides, Leda and John Toobey. "Evolutionary Psychology: A Primer." Accessed October 12, 2012, http://www.psych.ucsb.edu/research/cep/primer.html.

Cronbach, L. and R. Snow. *Aptitudes and Instructional Methods: A Handbook for Research on Interactions*. New York: Irvington, 1977.

"Cynicism." *In Our Time*. BBC Radio 4. October 20, 2005.

Damasio, Antonio. *The Feeling of What Happens: Body and Emotion in the Making of Consciousness*. New York: Mariner Books, 2000.

Damasio, Antonio. *Looking for Spinoza: Joy, Sorrow, and the Feeling Brain*. New York: Mariner Books, 2003.

Darwin, Charles. *The Expression of the Emotions in Man and Animals*. New York: D. Appleton, 1873.

Daus, C. S. "The Case for the Ability Based Model of Emotional Intelligence." In *A Critique of Emotional Intelligence: What Are the Problems and How Can They Be Fixed?* edited by Kevin Murphy. New York: Psychology Press, 2006.

De Sousa, Ronald. "The Rationality of Emotions." In *Explaining Emotions*, edited by Amélie Rorty. Oakland: University of California Press, 1980.

De Sousa, Ronald. *The Rationality of Emotions*. Cambridge: MIT Press, 1990.

Denis, Lara. "Freedom, Primacy, and Perfect Duties to Oneself." In *Kant's Metaphysics of Morals*, edited by Lara Denis. New York: Cambridge University Press, 2010.

Denis, Lara. "Kant's Conception of Virtue." In *The Cambridge Companion to Kant and Modern Philosophy*," edited by Paul Guyer. New York: Cambridge University Press, 2006.

Denis, Lara. *Moral Self-Regard: Duties to Oneself in Kant's Moral Theory*. New York: Routledge, 2001.

Dewey, John. "The Theory of Emotion. (Part II) The Significance of Emotions." *Psychological Review*, 2 (1895): 13–32.

Ehrenreich, Barbara. *Bright-Sided*. New York: Metropolitan Books, 2009.

Ekman, Paul. "Basic Emotions." In *Handbook of Cognition and Emotion*, edited by T. Dalgleish and M. Power. Sussex: John Wiley & Sons, 1999.

Ekman, Paul. *Emotion in the Human Face: Guidelines for Research and an Integration of Findings*. Oxford: Pergamon Press, 1972.

Ekman, Paul. "Expression and the Nature of Emotion." In *Approaches to Emotion*, edited by Klaus R. Sherer and Paul Ekman. Hillsdale, NJ: Lawrence Erlbaum, 1984.

Ekman, Paul. "A Methodological Discussion of Nonverbal Behavior." *Journal of Psychology*, 43 (1957): 141–149.

Engstrom, Stephen. "The Inner Freedom of Virtue." In *Kant's Metaphysics of Morals: Interpretive Essays*, edited by Mark Timmons. New York: Oxford University Press, 2002.

Engstrom, Stephen. "Reason, Desire, and the Will." In *Kant's Metaphysics of Morals*, edited by Lara Denis. New York: Cambridge University Press, 2010.

Epictetus. *Handbook of Epictetus*, translated by Nicholas P. White. Indianapolis: Hackett, 1983.

Feyerherm, A. E. and C. L. Rice. "Emotional Intelligence and Team Performance: The Good, the Bad, and the Ugly." *International Journal of Organizational Analysis*, 10 (2002): 343–362.

Fitness, J. "The Emotionally Intelligent Marriage." In *Emotional Intelligence in Everyday Life*, 2nd edition, edited by J. Ciarrochi, J. Forgas, and J. Mayer. New York: Psychology Press, 2006.

Flynn, James. "IQ Gains Over Time: Towards Finding the Causes." In *The Rising Curve: Long-Term Gains in IQ and Related Measures*, edited by U. Neisser. Washington, DC: American Psychological Association, 1998.

Fortenbaugh, W. W. *Aristotle on Emotion*. New York: Barnes & Noble Books, 1975.

Frankfurt, Harry. *The Importance of What We Care About*. Cambridge, UK: Cambridge University Press, 1987.

Freud, Sigmund. *Beyond the Pleasure Principle*. New York: W. W. Norton, 1990.

Freud, Sigmund. *Group Psychology and the Analysis of the Ego*. New York: W. W. Norton, 1990.

Freud, Sigmund. *The Psychopathology of Everyday Life*, translated by Anthea Bell. London: Penguin Books, 2002.

Frijda, Nico. *The Laws of Emotion*. London: Lawrence Erlbaum, 2007.

Gardner, Howard. *Intelligence Reframed: Multiple Intelligences for the 21st Century*. New York: Basic Books, 1999.

Geiger, Ido. "Rational Feeling and Moral Agency. " *Kantian Review*, 16 (2011): 283–308.

Gewirth, Alan. *Self-Fulfillment*. Princeton, NJ: Princeton University Press, 1998.

Giannini, Heidi Chamberlin. "Korsgaard and the *Wille/Willkür* Distinction: Radical Constructivism and the Imputability of Immoral Acts." *Kant Studies Online*, 72 (2013).

Goldie, Peter. *The Emotions: A Philosophical Exploration*. Oxford: Oxford University Press, 2002.

Goleman, Daniel. *Emotional Intelligence: Why It Can Matter More Than IQ*. New York: Bantam Books, 1995.

Goodman, Russell. "William James." In the *Stanford Encyclopedia of Philosophy*. Accessed September 26, 2014, http://plato.stanford.edu/entries/james/

Gould, Stephen J. *The Mismeasure of Man*. New York: W. W. Norton, 1996.

Greene, Joshua. *Moral Tribes*. New York: Penguin Press, 2013.

Greenspan, Patricia S. *Emotions and Reasons: An Inquiry into Emotional Justification*. New York: Routledge, 1988.

Gregor, Mary. *Laws of Freedom*. Oxford: Blackwell, 1963.

Grenberg, Jeanine. "What Is the Enemy of Virtue?" In *Kant's Metaphysics of Morals*, edited by Lara Denis. New York: Cambridge University Press, 2010.

Gressis, Robert. "How to Be Evil: The Moral Psychology of Immorality." In *The New Kant*, edited by Pablo Muchnik. Cambridge, UK: Cambridge Scholars, 2008.

Guyer, Paul. *Kant*. New York: Routledge, 2006.

Guyer, Paul. *Kant and the Experience of Freedom*. New York: Cambridge University Press, 1996.

Habermas, Jürgen. *Moral Consciousness and Communicative Action*, translated by Christian Lenhardt and Shierry Weber Nicholsen. Cambridge: MIT Press, 1995.

Habermas, Jürgen. *The Philosophical Discourse on Modernity*. Cambridge: MIT Press, 2000.

Halberstan, David. *The Children*. New York: Fawcett Books, 1999.

Hay, Carol. *Kantianism, Liberalism, and Feminism: Resisting Oppression*. New York: Palgrave Macmillan, 2013.

Heidegger, Martin. *Being and Time*, translated by John Macquarrie and Edward Robinson. New York: Harper Perennial, 1962.

Herman, Barbara. "Making Room for Character." In *Aristotle, Kant, and the Stoics: Rethinking Happiness and Duty*, edited by Stephen Engstrom and Jennifer Whiting. New York: Cambridge University Press, 1996.

Herman, Barbara. *The Practice of Moral Judgment*. Cambridge, MA: Harvard University Press, 2007.

Hochschild, A. R. *The Managed Heart: Commercialization of Human Feeling*. Berkeley: University of California Press, Berkeley, 1983.

Holland, Frederic. *The Reign of the Stoics*. New York: C. P. Somerby, 1879.

Holowchak, Andrew M. *The Stoics: A Guide for the Perplexed*. New York: Continuum, 2008.

Hookway, Christopher. "Pragmatism." In the *Stanford Encyclopedia of Philosophy*. Accessed September 26, 2014, http://plato.stanford.edu/entries/pragmatism/.

Horn, Christoph. "The Concept of Love in Kant's Virtue Ethics." In *Kant's Ethics of Virtue*, edited by Monika Betzler. Boston, MA: Walter De Gruyter, 2008.

Hume, David. *Treatise of Human Nature*. Oxford: Oxford Clarendon Press, 1888.

Hursthouse, Rosalind. *On Virtue Ethics*. New York: Oxford University Press, 1999.

Jagger, Alison. "Love and Knowledge: Emotion in Feminist Epistemology." In *Gender, Body, Knowledge*, edited by Alison Jagger and Susan R. Bordo. New Jersey: Rutgers University Press, 1989.

James, William. "What Is an Emotion." In *What Is an Emotion?* edited by Cheshire Calhoun and Robert Solomon. New York: Oxford University Press, 1984. Originally published in *Mind* (1884).

Janoff-Bulman, Ronnie. *Shattered Assumptions: Towards a New Psychology of Trauma*. New York: Free Press, 1992.

Johnson, Robert. *Self-Improvement*. New York: Oxford University Press, 2011.

Kane, Robert. *Through the Moral Maze*. New York: Parragon House, 1998.

Khamsi, Roxanne. "Impaired Emotional Processing Affects Moral Judgments." *NewScientist.com*, March 2007. Accesses September 29, 2014, http://www.newscientist.com/article/dn11433-impaired-emotional-processing-affects-moral-judgements.html#.VCl_aqvCS0Y

Klein, Jakob. "Of Archery and Virtue: Ancient and Modern Conceptions of Value." *Philosopher's Imprint*, 14 (19) (2014): 1–16.

Konstan, David. "The Emotion Is Aristotle Rhetoric 2.7." In *Influences on Peripatetic Rhetoric*, edited by David Mirhady. Amsterdam: Koninklijke Brill NV, 2007.

Kristjánsson, K. "'Emotional Intelligence' in the Classroom? An Aristotelian Technique." *Education Theory*, 56 (2006): 39–56.

Lazarus, Richard. *Emotion and Adaptation*. New York: Oxford University Press, 1991.

LeDoux, Joseph. *The Emotional Brain: The Mysterious Underpinnings of Emotional Life*. New York: Simon and Schuster, 1998.

Lewis, M. D. "Bridging Emotion Theory and Neurobiology through Dynamic System Modeling." *Behavioral and Brain Science*, 28 (2005): 105–131.

Lewis, M. D. "Self Organizing Cognitive Appraisals." *Cognition and Emotion*, 10 (1996): 1–26.

Lopes, Paulo N., John B. Nezlek, Astrid Schütz, and Peter Salovey. "Emotional Intelligence and Social Interaction." *Personality and Social Psychology Bulletin*, 30 (2004): 1018–1034.

Louden, Robert. "Anthropology from a Kantian Point of View: Toward a Cosmopolitan Conception of Human Nature." In *Rethinking Kant*, edited by Pablo Muchnik. Cambridge, UK: Cambridge Scholars, 2009.

Louden, Robert, "Evil Everywhere: The Ordinariness of Kantian Radical Evil." In *Kant's Anatomy of Evil*, edited by Sharon Anderson-Gold and Pablo Muchnik. New York: Cambridge University Press, 2010.

Louden, Robert. *Kant's Impure Ethics: From Rational Beings to Human Beings*. Oxford, New York: Oxford University Press, 2002.

Louden, Robert. *Morality and Moral Theory*. New York: Cambridge University Press, 1992.

MacBeath, A. Murray. "Kant on Moral Feeling," *Kant-Studien*, 64 (1973): 283–314.

MacCann, Carolyn, Gerald Matthews, Moshe Zeidner, and Richard D. Roberts. "The Assessment of Emotional Intelligence: On Frameworks, Fissures, and the Future." In *Measuring Emotional Intelligence: Common Ground and Controversy*, edited by Glenn Geher. New York: Nova Science, 2004.

Marcus, Robert F. "Emotion and Violence in Adolescence." In *Encyclopedia of Violence, Peace and Conflict,* 2nd edition, edited by Lester R. Kurtz. Waltham, MA: Academic Press, 2008.

Mayer, John D. and Peter Salovey. "What Is Emotional Intelligence?" In *Emotional Development and Emotional Intelligence: Educational Implications*, edited by Peter Salovey and David J. Sluyter. New York: Basic Books, 1997.

Mayer, John D. "A New Field Guide to Emotional Intelligence." In *Emotional Intelligence in Everyday Life*, edited by Joseph Ciarrochi, Joseph P. Forgas, and John D. Mayer. New York: Psychology Press, 2001.

Meerbote, Ralph. "*Wille* and *Willkür* in Kant's Theory of Action." In *Interpreting Kant*, edited by M. S. Gram. Iowa City: University of Iowa Press, 1982.

Muchnik, Pablo. *Kant's Theory of Evil: An Essay on the Dangers of Self-Love and the Aprioricity of History.* New York: Lexington Books, 2009.

Munzel, Felicitas G. *Kant's Conception of Moral Character.* Chicago: University of Chicago Press, 1999.

Murdoch, Iris. *The Sovereignty of the Good.* New York: Routledge, 2001.

Murphy, Kevin and L. Sideman. "What Is This Thing Called Emotional Intelligence?" In *A Critique of* Emotional Intelligence: What Are the Problems and How Can They Be Fixed? edited by Kevin Murphy. New York: Psychology Press, 2006.

Nell (O'Neill), Onora. *Acting on Principle.* New York: Columbia University Press, 1975.

Neiman, Susan. *Evil in Modern Thought.* Princeton, NJ: Princeton University Press, 2002.

Neu, Jerome. *A Tear Is an Intellectual Thing.* New York: Oxford University Press, 2002.

Nussbaum, Martha. *Upheavals of Thought: The Intelligence of Emotions.* New York: Cambridge University Press, 2001.

Olson, Lynne. *Freedom's Daughters: The Unsung Heroines of the Civil Rights Movement from 1830 to 1970.* New York: Scribner Press, 2001.

O'Neill, Onora. *Constructions of Reason.* Cambridge, UK: Cambridge University Press, 1990.

Packer, Mark. "Kant on Desire and Moral Pleasure." *Journal of the History of Ideas,* 50 (3) (1989): 429–442.

Paton, H. J. *The Categorical Imperative: A Study in Kant's Moral Philosophy.* Philadelphia: University of Pennsylvania Press, 1971.

Plomin, R., John C. DeFries, Gerald E. McClearn, and Peter McGuffin. *Behavioral Genetics,* 4th edition. New York: Freeman, 2001.

Prinz, Jesse. "Embodied Emotions." In *What Is an Emotion?* edited by Robert Solomon and Cheshire Calhoun. New York: Oxford University Press, 1984.

Prinz, Jesse. *The Emotional Construction of Morals.* New York: Oxford University Press, 2011.

Prinz, Jesse. *Gut Reaction: A Perceptual Theory of Emotion*. New York: Oxford University Press, 2004.

Purdon, Christine. "Thought Suppression and Psychopathology." *Behavior Research and Therapy*, 37 (11) (1999): 1029–1054.

Quine, W. V. O. "Two Dogmas of Empiricism." *The Philosophical Review*, 60 (1951): 20–43; reprinted in *From a Logical Point of View*. Cambridge, MA: Harvard University Press, 1953.

Rachels, James. *The Elements of Moral Philosophy*, 5th edition. New York: McGraw-Hill, 2006.

Rasmussen, Paul. *The Quest to Feel Good*. New York: Routledge Press, 2010.

Rawls, John. *Lectures on the History of Moral Philosophy*. Cambridge, MA: Harvard University Press, 2000.

Rawls, John. *A Theory of Justice*. Cambridge: Belknap Press, 1971.

Robinson, Jenefer. "Startle." *Journal of Philosophy*, 92 (1995): 53–57.

Rock, Irwin. "A Look Back at William James's Theory of Perception." In *Reflections on the Principles of Psychology: William James After a Century*, edited by Michael G. Johnson and Tracy B. Henley. Hillsdale, NJ: Lawrence Erlbaum Associates, 1990.

Rolls, Edmund T. *Emotion Explained*. New York: Oxford University Press, 2007.

Rorty, Amélie. "Explaining Emotions." *Journal of Philosophy*, 75 (3) (1978): 139–161.

Rorty, Amélie. "Explaining Emotions." In *Explaining Emotions*, edited by Amélie Rorty. Oakland: University of California Press, 1980.

Rorty, Amélie. "Introduction." In *Explaining Emotion*. Oakland: University of California Press, 1980.

Saarni, C. "Emotional Competence and Self-Regulation in Childhood." In *Emotional Development and Emotional Intelligence*, edited by P. Salovey and D. Sluyter. New York: Basic Books, 1997.

Salovey, Peter and John D. Mayer. "Emotional Intelligence." *Imagination, Cognition and Personality*, 9 (3) (1989–1990): 185–211.

Salovey, Peter and D. Sluyter, eds. *Emotional Development and Emotional Intelligence: Educational Implications*. New York: Basic Books, 1997.

Sandbach, F. H. *The Stoics*. Indianapolis: Hackett, 1994.

Schachter, S. and J. E. Singer. "Cognitive, Social, and Physiological Determinants of Emotional State." *Psychological Review*, 69 (5) (1962): 379–399.

Schnarch, David. *Passionate Marriage*. New York: W. W. Norton, 2009.

Schoda, Yuichi, Walter Mischel, and Philip K. Peake. "Predicting Adolescent Cognitive and Self-Regulatory Competencies From Preschool Delay Gratification." *Developmental Psychology*, 26 (6) (1990): 978–986.

Schulte, M. J., M. J. Ree, and T. R. Caretta. "Emotional Intelligence: Not Much More Than G and Personality." *Personality and Individual Differences*, 37 (2004): 1059–1068.

Schütz, Erica, Uta Sailer, Ali Al Nima, Patricia Rosenberg, Ann-Christine Andersson Arntén, Trevor Archer, and Danilo Garcia. "The Affective Profiles in the U.S.A.: Happiness, Depression, Life-Satisfaction, and Happiness-Increasing Strategies." *PeerJ* (2013). Accessed September 26, 2014, https://peerj.com/articles/156/.

Sedgwick, Sally. "Hegel's Critique of the Subjective Idealism of Kant's Ethics." *Journal of the History of Philosophy*, 29 (1) (1988): 89–105.

Setiya, Keiran. *Reasons without Rationalism*. Princeton, NJ: Princeton University Press, 2010.

Sherman, Nancy. *Making a Necessity of Virtue: Aristotle and Kant on Virtue*. New York: Cambridge University Press, 1997.

Sherman, Nancy. *Stoic Warriors*. New York: Oxford University Press, 2005.

Sherman, Nancy. *The Untold War: Inside the Hearts, Minds, and Souls of Our Soldiers*. New York: W. W. Norton, 2011.

Sidgwick, Henry. *The Methods of Ethics*. Indianapolis: Hackett, 1981.

Singer, Peter. *The Expanding Circle: Ethics and Sociology*. New York: Farrar, Straus, and Giroux, 1981.

Singer, Peter. *One World: The Ethics of Globalization*. Princeton, NJ: Princeton University Press, 2002.

Smith, Adam. *The Theory of Moral Sentiments*. Salt Lake City, UT: Gutenberg Press, 2011.

Sokoloff, William. "Kant and the Paradox of Respect." *American Journal of Political Science*, 45 (4) (2001): 768–779.

Sokolon, Marlene. *Political Emotion: Aristotle and the Symphony of Reason and Emotion*. DeKalb: Northern Illinois University Press, 2006.

Solomon, Robert. "Emotions and Choice." In *Explaining Emotions*, edited by Amélie Oksenberg Rorty. Oakland: University of California Press, 1980.

Solomon, Robert. "Emotions, Thoughts, and Feelings." In *Thinking about Feeling*, edited by Robert Solomon. New York: Oxford University Press, 2004.

Sorensen, Kelly. "Kant's Taxonomy of Emotions." *Kantian Review*, 6 (2002): 109–128.

Spearman, C. "'General Intelligence,' Objectively Determined and Measured." In *American Journal of Psychology*, 15 (1904): 201–293.

Spinoza, Baruch. *Ethics*, translated by Samuel Shirley, edited by Seymour Feldman. Indianapolis: Hackett, 1992.

Springer, Elise. *Communicating Moral Concern: An Ethics of Critical Responsiveness*. Cambridge: MIT Press, 2013.

Stevenson, C. L. *Ethics and Language*. New Haven, CT: Yale University Press, 1944.

Stevenson, Harold and James Stigler. *Learning Gap: Why Our Schools Are Failing and What We Can Learn from Japanese and Chinese Education*. New York: Touchstone, 1992.

Stocker, Michael and Elizabeth Hegeman. *Valuing Emotions*. New York: Cambridge University Press, 1996.

Sullivan, Roger. *Introduction to Kant's Metaphysics of Morals*. Cambridge, UK: Cambridge University Press, 1996.

Tennen, H. and G. Affleck, "Finding Benefits in Adversity." In *Coping: The Psychology of What Works*, edited by C. R. Snyder. New York: Oxford University Press, 1999.

Thorndike, Edward L. and S. Stein. "An Evaluation of the Attempts to Measure Social Intelligence." *Psychological Bulletin*, 34 (1937): 275–285.

Tough, Paul. *How Children Succeed: Grit, Curiosity, and the Hidden Power of Character*. Boston, MA: Houghton Mifflin Harcourt, 2012.

Tugade, M. M. and B. L. Fredrickson. "Positive Emotions and Emotional Intelligence." In *The Wisdom in Feeling: Psychological Processes in Emotional Intelligence*, edited by Lisa F. Barrett and Peter Salovey. New York: Guilford Press, 2002.

Urmson, J. O. "Saints and Heroes." In *Essays in Moral Theory*, edited by A. Melden. Seattle: University of Washington Press, 1958.

Velleman, J. David. *Practical Reflection*. New York: Oxford University Press,1989.

Walker, Margaret Urban. "Moral Understandings: Alternative Epistemology for a Feminist Ethics." *Hypatia*, 4 (2) (1989): 15–28.

Walker, Mark Thomas. *Kant, Schopenhaur, and Morality: Recovering the Categorical Imperative*. New York: Palgrave Macmillan Press, 2012.

Waterhouse, L. "Inadequate Evidence for Multiple Intelligences, Mozart Effect and Emotional Intelligence Theories." *Educational Psychologist*, 41 (2006): 247–255.

Williams, Bernard. *Moral Luck*. Cambridge, UK: Cambridge University Press, 1982.

Williamson, Diane. "Familial Duties and Emotional Intelligence." In *The Ethics of the Family*, edited by Stephen Scales, Adam Potthast, and Linda Oravecz. Cambrdige, UK: Cambridge Scholars, 2010.

Wolf, Susan. "Moral Saints." In *The Virtues*, edited by Robert B. Kruschwitz and Robert C. Roberts. Boston, MA: Wadsworth, 1987.

Wolf, Susan. *Freedom within Reason*. New York: Oxford University Press, 1990.

Wood, Allen. "The Final Form of Kant's Practical Philosophy." In *Kant's Metaphysics of Morals*, edited by Mark Timmons. New York: Oxford University Press, 2002.

Wood, Allen. "Kant and the Intelligibility of Evil." In *Kant's Anatomy of Evil*, edited by Sharon Anderson Gold and Pablo Muchnik. New York: Cambridge University Press, 2010.

Wood, Allen. "Kant's Compatibilism." In *Critical Essays on Kant's Critique of Pure Reason*, edited by Patricia Kitcher. New York: Rowman & Littlefield, 1998.

Wood, Allen. "Punishment, Retribution, and the Coercive Enforcement of Right." In *Kant's Metaphysics of Morals*, edited by Lara Denis. New York: Cambridge University Press, 2010.

Zeidner, M. and G. Matthews. "Personality and Intelligence." In *Handbook of Human Intelligence*, edited by R. J. Sternberg. Cambridge, UK: Cambridge University Press, 2000.

Zeidner, M., G. Matthews, and Richard D. Roberts, eds. *What We Know about Emotional Intelligence: How It Affects Learning, Work, Relationships, and Our Mental Health.* Cambridge: MIT Press, 2009.

Index

Printed in the United States
By Bookmasters